FM 3-92 (FM 100-15)

I0167870

Corps Operations

November 2010

Headquarters, Department of the Army

Foreword

The corps design was forged by Napoleon in the early 1800s and became the operational construct for decisive maneuver and exploitation through World Wars I and II, Korea, Panama, Operation Desert Storm, and Operation Iraqi Freedom.

Today, the corps remains the operational headquarters for decisive land combat and includes enhanced capabilities for full spectrum operations. The corps defines the fight, ensures coherency, conducts operational maneuver, and serves as the bridge to translate strategic guidance into tactical tasks. The corps serves in an essential role as a joint or multinational headquarters for many contingencies. It conducts contingency planning to shape the operational environment, execute decisive operations, integrate interagency and nongovernmental agency efforts, and assess operations. The staff translates the corps commander's visualization into plans and orders. Commanders consider the elements of operational design as they frame the problem and describe their visualization.

Field Manual 3-92 effectively describes the guiding principles and framework for decisive corps operations. It reflects the hard-earned gains and lessons learned during the conduct of corps operations in Iraq since 2003. The lessons learned continue to reinforce the absolute necessity of the corps as an operational headquarters. The corps operates in the temporal, physical, and functional realms between the tactical echelons and the strategic theater. It provides the unique capability to orchestrate large, complex operations and synchronize joint, multinational, and interagency actions in a coherent campaign for decisive full spectrum operations. It is through these efforts that tactical actions are linked to accomplish strategic campaign objectives.

Recent combat experience validates the enduring truth: the critical component in warfare at all echelons remains the leadership and professionalism of the Soldiers and supporting civilians that comprise the team. Inspired leadership, at the operational level provides the shared vision, purpose, and direction that guides this powerful capability.

As you study and reference this manual, bear in mind that the themes and concepts reflect the hard-earned experiences and sacrifices of units and Soldiers in combat in a tremendously complex operational environment against formidable and adaptable enemies. Apply these principles with relentless professionalism and even greater adaptability. We must prepare to fight our nation's wars, and the corps design provides the critical headquarters to orchestrate and conduct decisive, coherent, and aggressive joint and coalition operations to achieve operational objectives and strategic goals.

LTG LLOYD J. AUSTIN III
Commanding General
XVIII Airborne Corps, and Multi-National Corps – Iraq

Field Manual
No. 3-92 (100-15)

Headquarters
Department of the Army
Washington, D.C., 26 November 2010

Corps Operations

Contents

*This publication supersedes FM 100-15, 29 October 1996.

Figures

Tables

Preface

PURPOSE

This manual provides direction for the corps headquarters: what it looks like, how it is organized, how its staff operates, how it is commanded and controlled, and how it operates in full spectrum operations. This publication replaces the previous edition of the Army's corps operations manual and describes the organization and operations of the corps. It reflects current doctrine on the elements of full spectrum operations: offense, defense, and stability or civil support. This manual addresses these elements without regard to priority. This manual—

- Incorporates the Army's operational concept, full spectrum operations.
- Describes the stand alone corps headquarters.
- Links brigade combat team and division doctrine with theater army doctrine. It describes the principles underlying the Army modular corps.
- Shows a corps headquarters designed for four primary employment roles—in priority—an Army intermediate tactical headquarters, an ARFOR, a joint force land component command headquarters, and a joint task force headquarters.
- Recognizes that the corps headquarters normally requires augmentation with elements from theater-level organizations for selected missions.
- Discusses when serving as a joint task force or joint force land component command, the corps may require a separate subordinate headquarters to serve as the ARFOR or may need augmentation to serve as both the joint headquarters and the ARFOR.
- Introduces a staff organization that reflects the warfighting functions discussed in Field Manual (FM) 3-0.
- Describes the three designated command and control facilities: main command post, tactical command post, and mobile command group.
- Describes the ability of the corps headquarters to readily accept joint augmentation from a standing joint force headquarters core element or other joint manning and equipping source.

SCOPE

This publication is organized into five chapters and five appendixes:

- Chapter 1 introduces the modular corps headquarters concept.
- Chapter 2 discusses how the corps headquarters is organized.
- Chapter 3 describes corps command and control.
- Chapter 4 describes how the corps headquarters conducts full spectrum operations.
- Chapter 5 details how a corps headquarters transitions to the headquarters of a joint task force or joint force land component command.
- Appendix A explains sustainment in support of the corps headquarters.
- Appendix B discusses how the corps plans, prepares, executes, and assesses joint fires.
- Appendix C describes how the Army Network Enterprise Technology Command/9th Signal Command and other organizations support LandWarNet operations for the corps headquarters.
- Appendix D describes how the corps headquarters conducts airspace command and control.
- Appendix E describes Air Force planning considerations for corps operations.

APPLICABILITY

FM 3-92 applies to commanders and trainers at the corps echelon. It forms the foundation for corps operations curriculum within the Army school system.

FM 3-92 applies to the Active Army, the Army National Guard/Army/National Guard of the United States, and United States Army Reserve unless otherwise stated.

ADMINISTRATIVE INFORMATION

Most terms used in FM 3-92 that have joint or Army definitions are identified in the text. For terms defined in the text, the term is italicized before its definition, and the number of the proponent manual follows the definition. The glossary lists acronyms and abbreviations used in the text. Users must be familiar with Joint Publication (JP) 1-02 and FM 1-02, listed on page References-1.

This manual uses the phrase *corps forces* to indicate all organizations that have a command or support relationship to the corps assigned by a higher headquarters.

Headquarters, U.S. Army Training and Doctrine Command, is the proponent for this publication. The preparing agency is the Combined Arms Doctrine Directorate, U.S. Army Combined Arms Center. Send written comments and recommendations on a DA Form 2028 (*Recommended Changes to Publications and Blank Forms*) to Commander, U.S. Army Combined Arms Center and Fort Leavenworth, ATTN: ATZL-MCK-D (FM 3-92), 300 McPherson, Fort Leavenworth, KS 66027-2337; by e-mail to leav-cadd-web-cadd@conus.army.mil; or submit an electronic DA Form 2028.

Introduction

The United States Army has published doctrine on corps operations since the beginning of the twentieth century. Its experience with large unit operations began during the Civil War. Leaders on both sides realized that they could not command and control regiments, brigades, and divisions without an intermediate headquarters between the Army-level planning and supporting field operations and the lower echelon forces actually engaged in battle. Corps operations have been included in eleven Army field service regulations (and later field manuals) both for larger-unit operations spanning from 1905 to 1996. All reflected contemporary conditions and provided guidance to existing corps operations.

This manual describes the corps headquarters: a continental United States-based headquarters, with no assigned troops other than those in its headquarters battalion, which is deployable worldwide. The Army's two capstone publications, Field Manual (FM) 1 and FM 3-0, along with keystone publications FM 3-07 and FM 3-24, anticipate that future United States military operations will be joint campaigns requiring unity of effort by a team of military, civilian, joint, interagency, intergovernmental, and multinational organizations. The corps headquarters focuses on serving as an intermediate or senior tactical land echelon with the ability to command and control divisions, brigade combat teams, and associated functional and multifunctional support brigades. It remains capable of transitioning to a joint task force or joint force land component command headquarters.

The redesigned corps headquarters represents one of the biggest changes in Army organizations since World War II. Army of Excellence doctrine established the corps headquarters at the top of an organizational structure that contained thousands of Soldiers and numerous subordinate organizations. While they still exist in the Army forces structure, those subordinate forces are no longer assigned to the corps. The corps headquarters battalion contains the communications, life support, and command post elements to accomplish required tasks.

FM 3-92 reflects an Army corps headquarters designed to—in priority—command and control Army forces, command and control land components, and command and control joint forces for contingencies. Its primary mission is to command and control land forces in full spectrum operations. This manual discusses how the corps headquarters stays prepared to engage at any point across the spectrum of conflict as well as to command and control forces engaged in conditions of limited intervention, irregular warfare, and major combat operations.

This page intentionally left blank.

Chapter 1

The Corps

The Army's transformation to a brigade-based force has produced a modular, modified corps headquarters. This chapter introduces that organization and shows the corps headquarters in an operational environment. It explains the priorities of the redesigned corps headquarters able to exercise command and control over land forces or function as a joint task force.

THE JOINT ENVIRONMENT

1-1. The Army corps fights in a joint environment, whether subordinate to an Army or other Service headquarters. The corps is organized, trained, and employed to support the objectives of the joint force commander. The corps provides those command and control and warfighting capabilities that contribute to achieve unity of effort.

THE CORPS HEADQUARTERS

1-2. For the Army, the operational concept is full spectrum operations: Army forces combine offensive, defensive, and stability or civil support operations simultaneously as part of an interdependent joint force to seize, retain, and exploit the initiative, accepting prudent risk to create opportunities to achieve decisive results. They employ synchronized action—lethal and nonlethal—proportional to the mission and informed by a thorough understanding of all variables of the operational environment. Mission command that conveys the commander's intent and an appreciation of all aspects of the situation guides the adaptive use of Army forces. Full spectrum operations require continuous, simultaneous combinations of offensive, defensive, and stability or civil support tasks. (See Field Manual (FM) 3-0 for doctrine on the Army's operational concept.)

1-3. Full spectrum operations necessitate an expeditionary, scalable corps headquarters able to exercise command and control of land forces for operations (first priority) and be designated a joint task force headquarters (second priority). Each role grows increasingly complex and requires increasing amounts of augmentation. The theater army headquarters tailors the corps headquarters to meet mission requirements. The corps headquarters can exercise command and control over any combination of divisions, brigade combat teams, or support brigades provided from the Army pool of forces or other Service equivalents. It operates at either the tactical or the operational level of war, as required.

1-4. The corps headquarters has no organic troops other than its headquarters battalion. It routinely assumes attachment, operational control (OPCON), or tactical control (TACON) of numerous organizations. Until United States Army Forces Command has attached or given OPCON to units, the corps headquarters lacks training and readiness authority over division headquarters, brigade combat teams, and supporting brigades within the United States under the Army force generation process.

COMMAND AND CONTROL OF LAND FORCES

1-5. Exercising command and control of land forces for operations is the corps headquarters' first priority. If corps headquarters is not the senior Army headquarters within the area of operations, then it is an intermediate tactical headquarters. If it is the senior Army headquarters within a joint operations area (JOA), then it is the ARFOR for that joint task force (JTF) or multinational force. If the corps exercises command and control over all Marine Corps forces within the JOA, then it is a joint force land component command. If the corps also exercises command and control over multinational forces, then it is a coalition joint force land component command.

1-6. The corps headquarters exercises command and control of land forces for operations with little or no augmentation. When the mission dictates, the corps headquarters identifies and fills needs for specialized skills not organic to the corps headquarters. This is particularly true in prolonged stability operations. Requirements for augmentation increase with the complexity of the mission. When the corps acts as a joint force land component command, augmentation is provided according to the appropriate joint manning document.

Intermediate Tactical Headquarters

1-7. The corps headquarters acts as an intermediate tactical headquarters when conducting command and control of forces assigned, attached, or OPCON to it to conduct operations under a joint force land and component command with OPCON of multiple divisions (including multinational or Marine Corps formations) or other large tactical formations. This situation occurs when another corps or a field army (such as Eighth Army) is designated a joint force land and component command. Intermediate tactical headquarters use Army rather than joint tactics and procedures.

ARFOR

1-8. The senior tactical headquarters is the senior Army headquarters controlling multiple subordinate tactical formations. If a joint force land and component command is established, it is the senior tactical headquarters for that JTF headquarters. Identifying the senior tactical headquarters is significant because it receives an air support operations squadron from the Air Force. This squadron may be assigned to an intermediate tactical headquarters in multi-corps operations or when the geography separates units. Senior tactical headquarters function at the tactical and the operational levels of war simultaneously. They use Army rather than joint tactics and procedures.

1-9. When the corps headquarters is the senior tactical headquarters in a JOA, it is also the ARFOR for that JTF. The ARFOR includes both the senior Army headquarters controlling multiple subordinate tactical formations and the actual forces placed under a joint or multinational headquarters. The ARFOR is the Service component headquarters for a JTF or a joint and multinational force. The ARFOR commander answers to the Secretary of the Army through the Army Service component command for most administrative control or title 10 authorities and responsibilities (see FM 3-0, appendix B). As the senior Army headquarters, the corps provides administrative control for all Army units within the JTF, including those not under OPCON of the corps. The ARFOR may share some or all of its administrative control responsibilities with other Army headquarters based on the situation.

Joint Force Land Component Command

1-10. A corps headquarters designated by the JTF commander as a joint force land component command exercises command and control over all land force units in the JOA. Army units subordinate to it are normally OPCON, and Marine Corps forces assigned to it are normally under TACON. Multinational forces assigned to the joint force land component command form a combined joint force land component command if the nations are part of an alliance. If the multinational forces are an ad hoc grouping of nations, then it is a coalition joint forces land component command. The corps headquarters can transition to any joint force land component command with minimal joint augmentation. For sustained operations as a combined or coalition joint force land component command, the corps requires augmentation according to an appropriate joint manning document. Combined or coalition joint force land component commands are the senior tactical headquarters within a JTF headquarters. Typically, joint force land and component commands function at both the tactical and the operational levels of war simultaneously. Combined and coalition joint force land component commands use joint rather than Army tactics and procedures.

JOINT TASK FORCE

1-11. The second priority of the corps headquarters is to transition to a JTF headquarters. Minimum joint manning, defined as 20 other Service officers, is required for the headquarters to initiate operations as a JTF headquarters. This includes initiating campaign planning and deploying the corps' early entry command post and advance elements to establish initial command and control capabilities in the JOA. (See chapter 2 for a discussion of command posts.) The headquarters can also initiate shaping operations and coordinate with host-nation and multinational partners in the JOA. A corps headquarters acting as a JTF headquarters requires a separate ARFOR headquarters or significant augmentation because of the differing roles and responsibilities. JTFs focus on the operational level of war and use joint rather than Army tactics and procedures.

1-12. The Secretary of Defense or any joint force commander can establish a JTF. The size and scope of the force depends on the mission. A JTF can be established on a geographic or functional basis. Normally, the commander of the JTF exercises OPCON over forces and other resources the combatant commander allocates or apportions to the JTF. OPCON is the usual command relationship when the force conducts an operation with a limited objective that does not require centralized control of logistics. A JTF has Service components. It may also have subordinate JTFs or functional components. In a JTF organization, a corps can be a Service component, the headquarters of a subordinate JTF, or a functional organization. (See Joint Publication (JP) 1 for information on joint force organizations and interagency and intergovernmental coordination.)

AVAILABLE FORCES

1-13. Depending on the situation, the corps receives capabilities from theater army assets, depicted in figure 1-1, to support the operations. The theater army is a regionally focused command and control headquarters. The remaining chapters of this manual discuss the organization, command and control procedures, and operational activities of the corps headquarters.

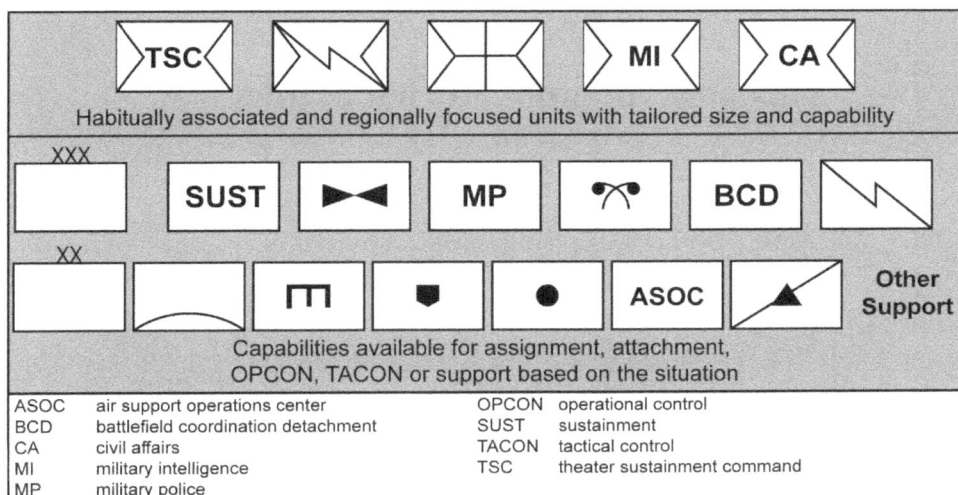

Figure 1-1. Forces available to support a corps

This page intentionally left blank.

Chapter 2

Corps Headquarters

Transformation has impacted the corps headquarters. This chapter describes the corps headquarters, its organization, the corps main and tactical command posts, and their facilities. It elaborates on how the Air Force supports the corps and describes the corps headquarters and headquarters battalion.

SECTION I – CORPS HEADQUARTERS ORGANIZATION

2-1. The corps headquarters is designed around several basic characteristics:

- When deployed, the headquarters is organized around one main command post (CP) and one tactical CP.
- The commanding general (CG) has a mobile command group so the CG can exercise command and control away from the CP. The mobile command group conducts some offensive and stability operations. The corps also fields an early-entry CP. This command and control element of the corps headquarters can control operations until remaining portions of the headquarters deploy. It conducts some operations—especially reception, staging, and onward movement and defensive operations. Normally, it locates near the tactical CP, and additional personnel from the main CP may augment it. Often, a deputy commander, assistant commander, chief of staff, or operations officer leads the early-entry CP. The corps headquarters battalion provides life support and network support to the headquarters. The corps tasks its subordinate units to provide security assets for each command and control facility. Alternatively multinational, host-nation, or contracted assets provide security. Regardless of its source, the corps headquarters security elements come under control of the corps headquarters battalion commander.
- The corps does not possess a set of corps troops other than those in its headquarters battalion. It receives attached forces. The corps exercises operational or tactical control over any mix of brigades, division headquarters, other Service, or multinational headquarters appropriate for its mission.

2-2. Figure 2-1 (page 2-2) depicts the corps headquarters organization consisting of a main CP, tactical CP, mobile command group, and its organic headquarters battalion with assigned companies. The corps main CP has three integrating and six functional cells, while the tactical CP forms a single integrating cell with functional components.

COMMAND GROUP

2-3. *Command* is the authority that a commander in the armed forces lawfully exercises over subordinates by virtue of rank or assignment. Command includes the authority and responsibility for effectively using available resources and for planning the employment of, organizing, directing, coordinating, and controlling military forces for the accomplishment of assigned missions. It also includes responsibility for health, welfare, morale, and discipline of assigned personnel (Joint Publication (JP) 1). The design of the corps identifies four general officer positions: the CG, deputy CG, chief of staff, and assistant chief of staff for operations. Paragraphs 2-4 through 2-10 discuss the duties of the two general officers with command responsibility, the CG and deputy CG. The duties of the chief of staff and assistant chief of staff for operations are addressed in the discussion of the staff.

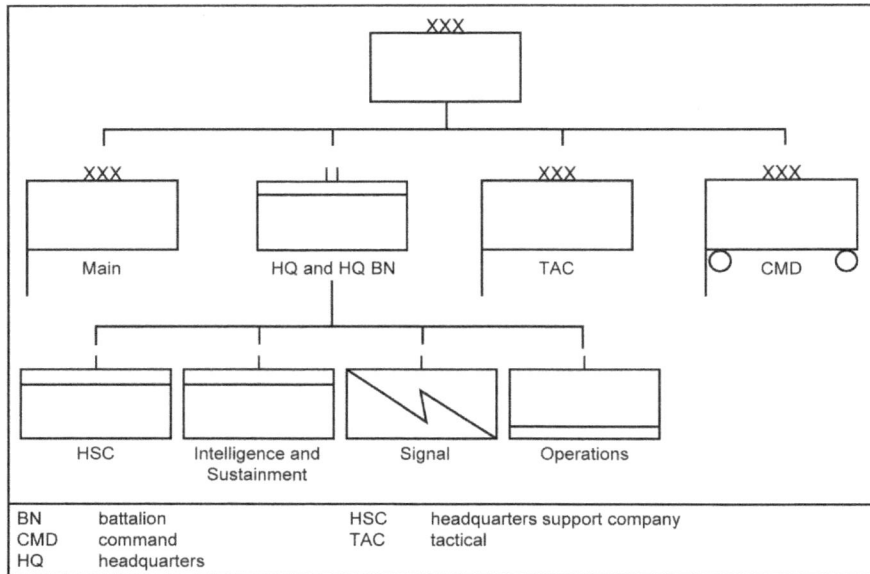

Figure 2-1. Corps headquarters and headquarters organization

COMMANDING GENERAL

2-4. A lieutenant general commands the corps. This CG has responsibility for the corps headquarters and all units and organizations that have a command or support relationship to the corps assigned by a higher headquarters (corps forces). The commander also oversees the control functions performed by the staff. Corps commanders normally position themselves where they can best exercise command and control over their corps.

DEPUTY COMMANDING GENERAL

2-5. The deputy CG, a major general, serves as the CG's primary assistant and second in command of the corps. The deputy CG has specific duties described in the corps standing operating procedures or directed by the commander. The deputy CG does not have a staff but may control certain staff elements based on responsibilities the CG assigns. The deputy CG can request staff assistance at any time.

2-6. The deputy CG interacts with the chief of staff and staff principal advisors based on duties the CG assigns. The deputy CG maintains situational understanding so to assume command at any time. Because of this requirement, the deputy CG normally remains at the main CP to be physically or virtually colocated with the CG. The deputy CG has three general responsibilities:

- Temporarily assume the CG's duties.
- Serve as the CG's successor.
- Assume certain delegated authorities.

Temporarily Assume the Commanding General's Duties

2-7. The deputy CG temporarily assumes command and makes decisions based on the mission and the commander's intent when needed. The deputy CG assumes the CG's duties temporarily on a routine basis during continuous operations for reasons of rest and health. A corps CG frequently leaves the main CP to meet with military and civilian superiors, government officials, and multinational partners. These meetings can take place outside the corps area of operations (AO), making it difficult for the CG to maintain a situational understanding of the current operation; therefore, the deputy CG steps in and assumes command during these events.

Serve as the Commanding General's Successor

2-8. The corps CG may be relieved, killed, wounded, or incapacitated in some way. In these situations, the deputy CG assumes command as specified in the standing operating procedures or the current order.

Assume Certain Delegated Authorities

2-9. The corps CG delegates authority to the deputy CG for specific tasks or areas. Such tasks can include the following:

- Provide control of reception, staging, onward movement, and integration activities.
- Coordinate directly with a host-nation partner.
- Provide command of multinational forces under corps control.
- Provide overall direction of sustainment activities.
- Conduct a specific shaping operation, such as a vertical envelopment or an amphibious operation.
- Conduct a decisive operation separated in time and space from the bulk of the corps assets.
- Serve as the chairman of any joint targeting or coordination boards established by the headquarters if corps serves as a base for a joint task force (JTF) or JTF headquarters.
- Supervise troops provided by the theater army that are not subordinated to a division.

2-10. Under certain conditions, the corps CG appoints a deputy commander from another Service or nation to demonstrate solidarity or to gain expertise in an area required for the operation. Such circumstances include when the corps serves as the base for a JTF, joint force land component command, or a multinational force. For example, a Marine Corps general officer may serve as the deputy CG for a corps engaging in amphibious operations.

THE STAFF

2-11. Staffs assist commanders in planning, coordinating, and supervising operations. A *staff section* is a grouping of staff members by area of expertise under a coordinating, special, or personal staff officer (Field Manual (FM) 5-0). Not all staff sections reside in one of the functional or integrating CP cells (discussed in paragraphs 2-24 through 2-32). These staff sections maintain their distinct organizations. They operate in different CP cells as required and coordinate their activities in meetings to include working groups and boards established by the unit's battle rhythm. *Battle rhythm* refers to a deliberate daily cycle of command, staff, and unit activities intended to synchronize current and future operations (JP 3-33). The staff consists of the chief of staff, personal staff, special staff, coordinating staff, staff augmentation, and command liaison, functional liaison, and others. See chapter 3 for a discussion of battle rhythm at corps level.

Chief of Staff

2-12. The corps CG delegates supervision of the staff to the chief of staff. The chief of staff directs, supervises, and trains the staff and is one of the CG's principal advisors. An effective working relationship helps the chief of staff and the CG transmit and share information and insights.

2-13. All staff principal advisors report to the chief of staff. The chief of staff is the command's principal integrator. This duty includes overseeing the command and control functional cell and its components of civil affairs operations, psychological operations, information engagement, and network operations. The chief of staff normally remains at the main CP. (FM 6-0 lists the chief of staff's responsibilities.)

Personal Staff

2-14. The personal staff sections advise the commander, provide input to orders and plans, and interface and coordinate with entities external to the corps headquarters. They perform special assignments as directed by the commander. Army regulations and public law establish special relationships between certain staff officers and the commander. For example, Army Regulation (AR) 20-1, AR 27-1, AR 165-1, and AR 360-1 require the inspector general, staff judge advocate, chaplain, and public affairs officer to be members of the commander's personal staff. Additionally, the command sergeant major and aides are part of the commander's personal staff. (FM 6-0 discusses the duties of the personal staff.)

Special Staff

2-15. Special staff officers help commanders and other staff members perform their responsibilities. The number of special staff officers and their duties vary. Special staff sections are organized according to professional or technical responsibilities. The commander delegates planning and supervisory authority over each special staff function to a coordinating staff officer. Although special staff sections may not be integral to a coordinating staff section, they usually share areas of common interest and habitual association. Special staff officers routinely deal with more than one coordinating staff officer. (FM 6-0 details the types and responsibilities of special staff officers.) The members of the special staff can change depending on the capabilities available to the corps commander and on the situation. (Table 2-1 identifies the most common special staff members.)

Table 2-1. Personal and special staff officers

Personal Staff	Special Staff
• Staff Judge Advocate • Public affairs officer • Political advisor • Inspector General • Chaplain	• Air and missile defense officer • Aviation officer • Air mobility liaison officer • Air liaison officer • CBRN officer • Engineer officer • Electronic warfare officer • Historian • Knowledge management officer • Operations research/systems analysis officer • Provost marshal • Red team officer • Safety officer • Staff weather officer • Transportation officer
CBRN chemical, biological, radiological, and nuclear	

Coordinating Staff

2-16. The corps coordinating staff officers are assistant chiefs of staff. They report to the chief of staff and have functional responsibilities in addition to their roles in the functional and integrating cells in the main and tactical CPs. Coordinating staff officers advise, plan, and coordinate actions within their areas of expertise. They also exercise planning and supervisory authority over designated special staff officers as described in FM 6-0.

Staff Augmentation

2-17. Depending on the situation, teams and detachments with special expertise to facilitate mission accomplishment augment the corps headquarters. Available capabilities include civil affairs operations, space support, combat camera, operational law, internment and resettlement, history, and public health.

COMMAND LIAISON, FUNCTIONAL LIAISON, AND OTHERS

2-18. These organizations vary in size with staff depending on the situation. The corps and higher, lower, and subordinate headquarters exchange command liaison teams. Host-nation and other nonmilitary entities also exchange command liaison teams. Functional liaison teams work with those organizations that provide services to the corps such as intelligence, signal, and sustainment. The corps headquarters selects, trains, and equips liaison teams from the corps staff for their responsibilities. The corps headquarters receives, houses, provides life support, and trains liaison teams from outside the corps headquarters to do their job properly.

COMMAND POST CELLS AND OTHER COMMAND POST ORGANIZATIONS

2-19. A *command and control system* is the arrangement of personnel, information management, procedures, and equipment and facilities essential for the commander to conduct operations (FM 6-0). The primary facilities of any command and control system are its command posts. A *command post* is a unit headquarters where the commander and staff perform their activities (FM 5-0). Organizing the staff into CPs expands the commander's ability to exercise command and control and makes the command and control system more survivable. Organizing the command posts into functional cells and integrating cells facilitates cross-functional coordination, synchronization, and information sharing.

2-20. For most operations, the corps headquarters operates from its main CP. If the main CP is not fully functional, the tactical CP deploys. The main CP usually does not displace during operations. However, sometimes it may have to relocate. One example is a move from an intermediate staging base to the AO. Another is a situation where political pressure dictates a move to another location.

2-21. A *command post cell* is a grouping of personnel and equipment by warfighting function or planning horizon to facilitate the exercise of command and control (FM 5-0). Two types of CP cells exist, functional and integrating. Functional cells group personnel and equipment by warfighting function. (FM 3-0 discusses the warfighting functions.) Integrating cells group personnel and equipment to integrate functional cell activities. Integrating cells normally focus on different planning horizons. All staffs sections and CP cells—coordinating, functional, integrating, special, and personal—integrate information and activities for mission accomplishment.

2-22. The corps headquarters design modifies the organization of the corps headquarters around the traditional, special, or coordinating staff concept. The traditional concept uses the naming conventions denoted in table 2-2 (page 2-6). Members of the coordinating staff serve in the functional and integrating cells. They are always members of their coordinating staff section; however, during operations they may simultaneously be a part of a functional cell and a part of the integrating cell.

Table 2-2. Naming conventions for staff officers in corps

G-1	assistant chief of staff, personnel
G-2	assistant chief of staff, intelligence
G-3	assistant chief of staff, operations
G-4	assistant chief of staff, logistics
G-5	assistant chief of staff, plans
G-6	assistant chief of staff, signal
G-7	assistant chief of staff, information engagement
G-8	assistant chief of staff, financial management
G-9	assistant chief of staff, civil affairs operations

2-23. Within a CP, commanders normally organize their corps headquarters staff into six functional cells, three integrating cells, and other groupings (meetings) as needed. Some corps staff members are permanently assigned to and serve one cell or element for the mission. However, others move from cell to cell based on the need for their expertise. It is possible for one member of the corps staff to serve in five or more components of the main or tactical CP depending on that member's skill set and ability to contribute to the corps mission. Table 2-3 lists the multiple positions a single colonel serving in the corps might have to fill. Similar multitasking often occurs for senior members of the corps headquarters. This especially applies for high-value, low-density members of the corps headquarters, such as chaplains, the political-military advisor, and interpreters or translators. (Figure 2-2, page 2-8, shows a main CP organized into CP cells containing staff elements and other groupings.)

Table 2-3. Example of positions a corps colonel will fill in command post activities

Activity	Position
Ad hoc grouping	Human resources policy board chair
Special staff	Main command post safety officer
Integrating cell	Plans cell replacement policy subject matter expert
Functional cell	Main command post sustainment cell deputy officer in charge
Coordinating staff	Corps assistant chief of staff, G-1/AG (S-1), personnel

FUNCTIONAL CELLS

2-24. The six functional cells each contribute to the development and maintenance of the corps's command and control capabilities. Generally, the title of each functional cell describes the cell: movement and maneuver, intelligence, fires, sustainment, command and control, and protection. Members of each cell specialize in activities related to that function.

2-25. Not all functional cells are permanently represented in all the integrating cells. They provide representation as required. For the staff principal advisors, this arrangement requires focus and discipline to maintain a suitable span of control.

2-26. Personal and some coordinating and special staff officers do not reside in a CP functional cell. They maintain a separate identity to address the areas of their special expertise. Nonetheless, they and their sections interact continuously with other staff sections. Officers for civil affairs operations and information engagement (see paragraph 2-116 for the definition of information engagement) and the political advisor may contribute expertise to a civil military operations center or similar organization to provide focus on their specialty. These staff sections maintain their distinct organization and operate in different CP cells as required. They coordinate their activities in the various meetings (including boards and working groups) identified in the unit's battle rhythm. (FM 5-0 discusses battle rhythm further.)

INTEGRATING CELLS

2-27. The corps headquarters has three integrating cells: current operations integration, future operations, and plans. Integrating involves coordinating or unifying activities across functions. Integrating cells group Soldiers and equipment to integrate the warfighting functions by planning horizon. A *planning horizon* is point in time commanders use to focus the organization's planning efforts to shape future events (FM 5-0).

2-28. Individual members of personal and special staff sections support the integrating cells as needed. These staff section members represent their sections in addition to supporting the cells.

2-29. During operations, the lines dividing the current operations integration, future operations, and plans cells often overlap. All people in the corps headquarters take part in planning, regardless of their positions or titles. The flexibility required in today's operations means that the plans, future operations, and current operations integration cells sometimes all work on different planning horizons of the same operation. Further, integration also occurs in the CP functional cells as they work internally and between cells to solve the problems presented during operations.

Current Operations Integration Cell

2-30. The current operations integration cell oversees day-to-day operations. It executes tactical operations and decisionmaking, including maintaining status and conducting update briefings. The current operations integration cell is a vital element of the corps main headquarters. Other current operations integration cell elements from the corps staff support it.

Future Operations Cell

2-31. The future operations cell performs mid-range planning, including preparation of branches to the current operation. The cell tracks and processes relevant information to create an ongoing link between current operations and plans. The future operations cell links the current operations integration and plans cells. As required, it augments them.

Plans Cell

2-32. The plans cell performs long-range planning. It develops complete operation plans or operation orders and sequels that the staff formally passes to the future operations or current operations cells, as required, for additional planning or execution.

MEETINGS, INCLUDING BOARDS AND WORKING GROUPS

2-33. The corps CG and chief of staff establish meetings, including boards and working groups, to further integrate the staff and enhance planning and decisionmaking. The CG uses boards and working groups, such as the targeting board and assessment working group, as the situation requires. Boards and working groups are established, modified, and dissolved as the situation evolves. The chief of staff manages the timings of these events through the corps battle rhythm. The chief of staff uses the battle rhythm to sequence command and control activities within a headquarters and throughout the force to facilitate effective command and control. Some meetings convene daily at a set time; others meet on call to address occasional requirements. The CG also identifies staff members to participate in the higher headquarters' boards and working groups. (JP 3-31, JP 3-33, and JP 4-0 discuss boards and working groups used by JTF and joint force land component command staffs respectively.)

SECTION II – MAIN COMMAND POST

2-34. The main CP is organized around six functional cells and three integrating cells. It synchronizes the conduct of current operations and allocates available resources. Under the general supervision of the corps chief of staff, it also oversees the conduct of future planning, analysis for current and future operations, sustainment coordination, and other staff functions.

2-35. The main CP operates from a fixed location. It normally does not displace during execution of operations. The main CP has coordinating, personal, and special staffs. The primary components of the main CP are the command group, the headquarters battalion, and the functional and integrating cells. (See figure 2-2.) The command group consists of the personal and special staff, command liaison, and functional liaison. These assets enable the commander to exercise command and control and show command presence away from a CP. Soldiers in the main and tactical CPs and other command and control facilities are assigned to the headquarters battalion. In addition to the life and personal support functions, the headquarters battalion enables the CP to support the exercise of command and control by the commander and staff.

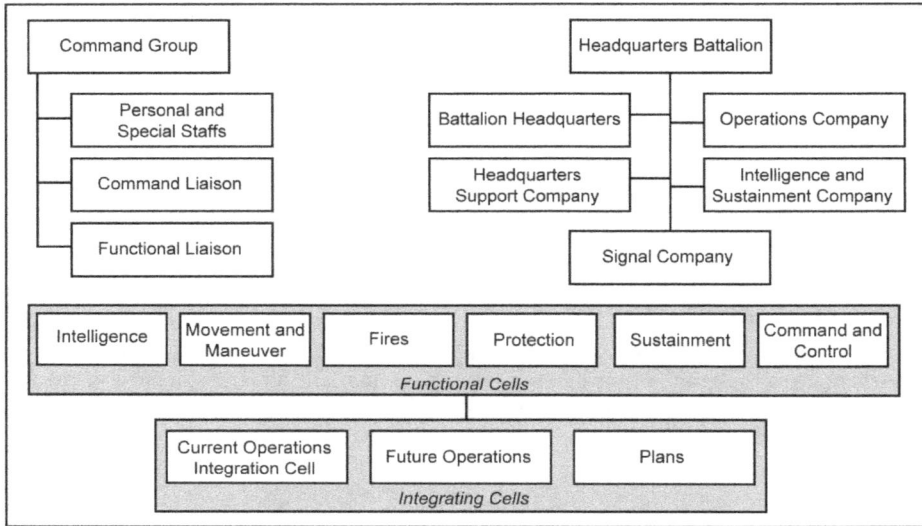

Figure 2-2. Main command post

COORDINATING, SPECIAL, AND PERSONAL STAFF OFFICERS

2-36. Table 2-4 (page 2-9) lists the senior leaders in the main CP. They occupy the duty locations indicated unless directed elsewhere. As currently designed, the coordinating, special, and personal staff officers of the corps operate in several environments simultaneously, each with slightly different responsibilities.

Table 2-4. Main command post staff principal advisors

Title	Grade	Cell
Commanding general	O-9	command group
Deputy commanding general	O-8	command group
Chief of staff	O-8	command group
Assistant chief of staff, G-1/AG, personnel	O-6	sustainment
Assistant chief of staff, G-2, intelligence	O-6	intelligence, surveillance, and reconnaissance officer in charge
Assistant chief of staff, G-3, operations	O-7	movement and maneuver officer in charge
Assistant chief of staff, G-4, logistics	O-6	sustainment officer in charge
Assistant chief of staff, G-5, plans	O-6	plans officer in charge
Assistant chief of staff, G-6, signal	O-6	command and control
Assistant chief of staff, G-7, information engagement	O-6	command and control
Assistant chief of staff, G-8, financial management	O-6	sustainment
Surgeon	O-6	sustainment
Assistant chief of staff, G-9, civil affairs operations	O-6	command and control
Chemical, biological, radiological, and nuclear officer	O-6	protection
Chaplain	O-6	personal staff
Cultural advisor	Civ	personal staff
Engineer coordinator	O-6	movement and maneuver
Chief of fires	O-6	fires officer in charge
Air liaison officer	O-6	special staff
Inspector general	O-6	personal staff
Command liaison officer	O-6	personal staff
Operations research/systems analysis officer	O-5	special staff
Political advisor	Civ	personal staff
Chief of protection	O-6	protection officer in charge
Provost marshal	O-6	protection cell
Public affairs officer	O-6	personal staff
Red team officer	O-6	special staff
Space operations officer	O-5	movement and maneuver
Staff Judge Advocate	O-6	personal staff

2-37. Staffing the tactical CP mirrors that of the main CP with the exception of grade structure. The tactical CP focuses on the conduct of specific current operations. Table 2-5 (page 2-10) lists the current staff principal advisors in the tactical CP. The mission and situation determine the officer in charge of the tactical CP. For example, a tactical CP controls a shaping operation and coordinates with a host-nation government or armed forces while the main CP exercises command and control of decisive operations. In this case, the corps deputy CG becomes the officer in charge to permit general officer-to-general officer discussion. Conversely, use of the corps tactical CP to monitor reception, staging, onward movement, and integration at a seaport of embarkation requires an officer of lesser rank.

Table 2-5. Tactical command post staff principal advisors

Title	Grade	Cell
Tactical command post officer in charge	-	as designated
Assistant G-2	O-5	intelligence
Deputy G-3, chief of operations	O-5	movement and maneuver officer in charge
Airspace command and control officer	O-4	movement and maneuver
Aviation officer	O-4	movement and maneuver
Engineer officer	O-4	movement and maneuver
Judge advocate officer	O-4	movement and maneuver
CBRN officer	O-4	movement and maneuver
Civil affairs operations chief	O-5	command and control
Deputy chief of fires	O-5	fires officer in charge
Air liaison officer	O-5	fires
Sustainment operations officer	O-5	sustainment officer in charge
Deputy G-6	O-5	command and control
Provost marshal	O-4	protection officer in charge

2-38. The commander and staff receive support from the headquarters battalion through its four subordinate companies and its battalion staff. The companies and staff provide administrative, logistic, life, and transportation support for all organic elements of the corps headquarters in garrison and when deployed. With teams identified for the main and tactical CPs, the headquarters battalion supplies unit-level command and control, communications, transportation, medical, food service, and maintenance support for the command and control nodes.

MAIN COMMAND POST FUNCTIONAL CELLS

2-39. The main CP's six functional cells coordinate and synchronize forces and activities by warfighting function. The six functional cells have assigned personnel in the movement and maneuver current operations integration cell to synchronize staff functions daily. These personnel are depicted as current operations integration cell (in figures 2-3 through 2-8) under their respective functional discussions. These current operations integration cell support elements from functional cells are an important part of the main CP. Some current operations integration cell support members may come from personal and special staff sections. The functional cells within a corps CP are—

- Intelligence.
- Movement and maneuver.
- Fires.
- Protection.
- Sustainment.
- Command and control.

INTELLIGENCE CELL

2-40. The intelligence functional cell is concerned with facilitating understanding of the operational environment. The cell requests, receives, and analyzes information from all sources. It disseminates intelligence products to support corps operations and the CG's situational understanding. This cell manages the requirements for all collection assets under corps control. The assistant chief of staff, intelligence (G-2) is normally the cell chief. The intelligence cell has three principal sections: intelligence, surveillance, and reconnaissance (ISR) operations section; G-2 analysis control element; and counterintelligence and human intelligence operations section (G-2X). (See figure 2-3.) The intelligence cell provides representatives to the current operations integration cell.

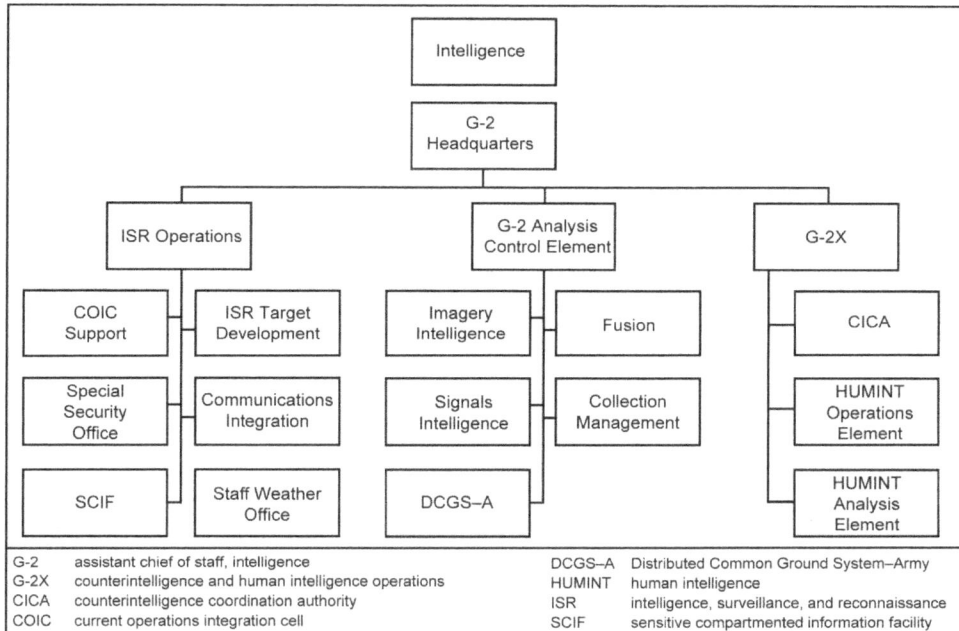

G-2	assistant chief of staff, intelligence	DCGS–A	Distributed Common Ground System–Army
G-2X	counterintelligence and human intelligence operations	HUMINT	human intelligence
CICA	counterintelligence coordination authority	ISR	intelligence, surveillance, and reconnaissance
COIC	current operations integration cell	SCIF	sensitive compartmented information facility

Figure 2-3. Main command post intelligence cell

2-41. To support corps operations, the main CP intelligence cell—

- Receives, processes, and analyzes information from all sources and disseminates intelligence.
- Provides relevant intelligence to support current and future operations activities.
- Synchronizes and integrates ISR operations.
- Participates in the targeting process.
- Conducts intelligence collection management, including planning, synchronizing, and integrating assets.
- Plans, monitors, and analyzes human intelligence and counterintelligence activities.

Intelligence, Surveillance, and Reconnaissance Operations

2-42. This section serves as the operations hub for ISR activities. It interfaces the intelligence cell with the movement and maneuver cell to integrate intelligence products and collection activities into current operations. It recommends ISR tasks to the senior intelligence officer for resources under corps control. This section receives, processes, analyzes, and disseminates all-source intelligence to support current and future operations. (See chapter 4 for more information about ISR.)

Current Operations Integration Cell Support

2-43. This element provides intelligence capability to the main CP through the integration of intelligence products and collection planning. Personnel in this element support the main CP current operations integration cell.

Special Security Office and Sensitive Compartmented Information Facility

2-44. The special security office exercises oversight of sensitive compartmented information reception, transmission, and storage. This office establishes, manages, and provides security for the corps main sensitive compartmented information facility. This facility, an accredited area with personnel access control, stores sensitive compartmented information. Personnel can also use, discuss, and process this information in this facility.

Intelligence, Surveillance, and Reconnaissance Target Development

2-45. This element develops and nominates priority targets as part of the targeting process. Paragraph B-8 discusses targeting in more detail. (See JP 3-60 for additional information on the joint targeting process.)

Communications Integration

2-46. This element establishes communications connectivity with outside intelligence elements, maintains internal and external digital communications functions, and exercises communications security oversight.

Staff Weather Office

2-47. This element, staffed by Air Force personnel, provides staff weather, forecasting, and observation support to the corps commander and staff.

G-2 Analysis Control Element

2-48. The G-2 analysis control element performs collection management, produces all-source intelligence, provides intelligence and electronic warfare technical control, and disseminates intelligence and targeting data across the spectrum of conflict.

Imagery Intelligence

2-49. This element serves as the single-source intelligence point of contact for exploitation and analysis of imagery and development of imagery products.

Signals Intelligence

2-50. This element performs signals intelligence analysis, electronic intelligence preparation of the battlefield, and tasks signals intelligence systems to support the ongoing operation.

Distributed Common Ground System-Army: Tactical Exploitation System–Forward

2-51. This element receives, processes, exploits, and disseminates signals intelligence, imagery intelligence, measurement and signature intelligence, and geospatial information and products. The end result produces multisource products to support current operations.

Fusion

2-52. This element performs situation development, prepares combat assessments, and develops and updates threat information for the ongoing and intelligence running estimate.

Collection Management

2-53. This element monitors collection asset status, develops the collection plan, and integrates and synchronizes assets to optimize intelligence collection. It focuses the employment of collection assets to satisfy the commander's priority intelligence requirements and information requirements.

Counterintelligence and Human Intelligence

2-54. This section advises the CG and the senior intelligence officer on employing counterintelligence and human intelligence assets. It interfaces with external sources to synchronize and deconflict counterintelligence and human intelligence operations.

Counterintelligence Coordination Authority

2-55. This element provides technical control, oversight, and deconfliction for counterintelligence assets.

Human Intelligence Operations

2-56. This element provides primary technical control and deconfliction for all human intelligence assets in the corps AO.

Human Intelligence Analysis

2-57. This element serves as the single fusion point for human intelligence reporting and operational analysis. It answers requests for information related to human intelligence.

MOVEMENT AND MANEUVER CELL

2-58. The movement and maneuver functional cell contains sections concerned with moving forces to achieve a position of advantage in relation to the enemy. The assistant chief of staff, operations (G-3) is the chief of the movement and maneuver cell. This functional cell forms the base of the current operations integration, future operations, and plans integrating cells. The current operations integration cell includes the elements shown in figure 2-4 (page 2-14). The G-3 exercises staff supervision over the integrating cells consistent with the guidance and oversight of the chief of staff.

2-59. For the corps headquarters, the main CP movement and maneuver cell—
- Oversees reception, staging, onward movement, and integration operations.
- Conducts force positioning and maneuver.
- Prepares orders and plans, including branches and sequels.
- Monitors current operations, maintains ongoing operations, and communicates status information throughout the command.
- Provides airspace management and deconfliction.
- Coordinates and synchronizes aviation operations.
- Coordinates and synchronizes space support.
- Provides terrain visualization and terrain products.
- Coordinates combat engineering, general engineering, and geospatial engineering.
- Provides liaison to and from subordinate, lateral, and higher headquarters.

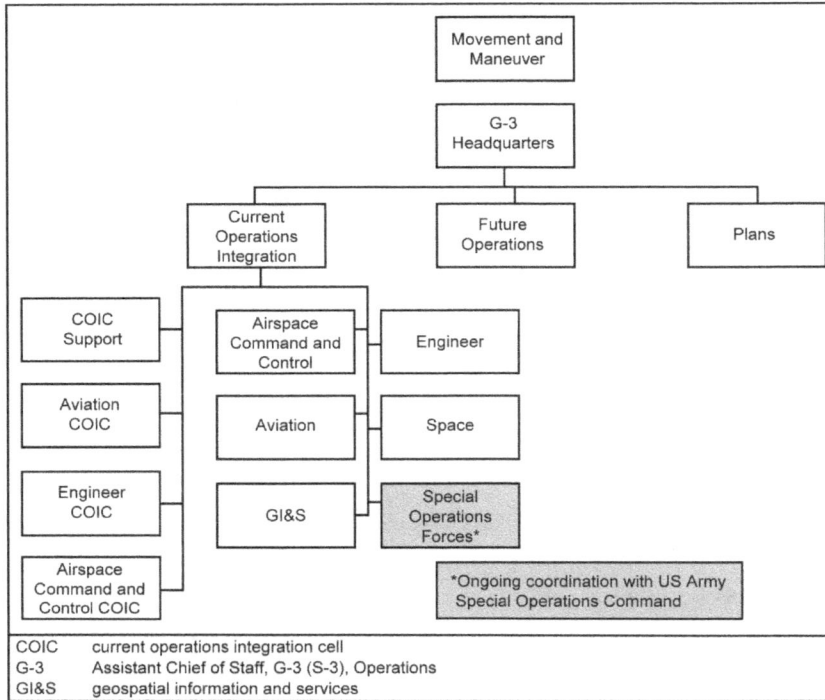

Figure 2-4. Main command post movement and maneuver cell

Current Operations Integration Cell

2-60. The current operations integration cell is the hub of the main CP. The current operations integration cell conducts the day-to-day activities within the corps. See paragraphs 2-129 through 2-131 for more information about current operations integration cells.

Current Operations Integration Cell Support

2-61. This movement and maneuver cell forms the core of the current operations integration cell. The chief of operations, a colonel, has responsibility for synchronizing all the current operations integration cell support sections. Some important functions are executing the tactical operations for the main CP, executing decisionmaking for the main CP, and conducting CP operations.

Aviation Current Operations Integration Cell Support

2-62. This section supports the coordination and synchronization of operational and tactical aviation maneuver support for the corps within the current operation integration cell.

Engineer Current Operations Integration Cell Support

2-63. This section within the current operations integration cell supports the functions of all engineers assigned and attached to the corps for operations

Airspace Command and Control Current Operations Integration Cell Support

2-64. This section within the current operations integration cell supports the airspace management and deconfliction for the corps.

Airspace Command and Control Section

2-65. The airspace command and control section provides airspace management in the corps AO. It provides input to the airspace control plan developed by the airspace control authority. The airspace command and control section develops standing operating procedures and AC2 annexes that facilitate standardized AC2 operations among subordinate units. These standing operating procedures and annexes align with joint airspace and theater army CP procedures, the appropriate aeronautical information publication, and associated plans and orders. For additional information, see appendix D.

Aviation Section

2-66. The aviation section coordinates and synchronizes the execution of operational and tactical aviation maneuver and maneuver support and aviation maneuver sustainment operations. It also coordinates and synchronizes unmanned aerial reconnaissance, close combat attack, mobile strike, vertical envelopment, air assault, battle command of the move, medical evacuation, and transportation of key personnel.

Geospatial Information and Services Section

2-67. Geospatial intelligence is an intelligence discipline that draws on contributions from both the intelligence and engineer communities to exploit imagery and geospatial information in describing the operational environment's effects on enemy and friendly capabilities. The geospatial information and services section acquires, manages, and distributes geospatial data and terrain visualization products to the CG and staff. This section includes imagery analysts and geospatial engineers from the corps' organic geospatial engineer team and may include augmentation from the National Geospatial-Intelligence Agency. See JP 2-03 for more information.

Engineer Section

2-68. This section includes the corps engineer who advises the commander and staff on engineering and the use of engineering assets. As the primary engineer section within the corps staff, it typically includes the senior engineer on the staff, the engineer coordinator. The engineer coordinator coordinates engineer tasks related to combat, general and geospatial engineering facilitating the functions which assure mobility, enhance protection, enable expeditionary logistics, and facilitate capacity building. The engineer coordinator provides guidance and reachback for the engineer section in the main CP. This section also coordinates with the engineer element in the protection cell to address specific engineer support for preserving the force such as base camp development planning. The engineer section coordinates and synchronizes engineer operations within the corps and with other headquarters, between echelons, and with multinational forces, governmental, and nongovernmental organizations. See FM 3-34 for more information.

Space Section

2-69. The space section is the CG's primary planner and advisor for space capabilities. To support corps operations, it maintains space situational awareness and coordinates with higher headquarters space elements, the Army space coordination section, the space coordinating authority staff, and managers of space-based systems, including the director of space forces. The space section serves as the primary coordinating element for the corps with—

- Space operations.
- Special technical operations.
- Alternative or compensatory control measures.

Special Operations Forces Section

2-70. The special operations coordination section oversees ongoing coordination between the corps and the Army special operations command. For more detail, see FM 3-05.

Future Operations Cell

2-71. The future operations cell is an important bridge between current operations and plans. See paragraphs 2-132 through 2-134 for more information about the future operations cell.

Plans Cell

2-72. The plans cell conducts planning for the corps. See paragraphs 2-137 through 2-140 for more information about the plans cell.

FIRES CELL

2-73. The fires functional cell coordinates Army indirect fires, joint fires, and command and control warfare, including nonlethal actions, through the targeting process. The cell implements the commander's intent by destroying enemy warfighting capabilities, applying nonlethal actions, and degrading enemy command and control capabilities through command and control warfare. The fires cell accomplishes these actions by developing, recommending, and briefing the scheme of fires, including both lethal fires and nonlethal actions (electronic attack and computer network operations with the effects of other warfighting functions) to the commander.

2-74. The corps chief of fires is a coordinating staff officer who leads the corps fire cell at the corps main CP. The chief of fires may locate at the tactical CP or elsewhere depending on the situation. See appendix B and FM 6-0 for further discussion of the chief of fires.

2-75. When the corps serves as the headquarters for a JTF or a joint force land component command, the fires cell performs additional functions. See JP 3-09 and JP 3-31.

2-76. The fires functional cell works closely with the force field artillery headquarters, if one is established. The fires cell's responsibilities are based on the situation and may include coordination and technical oversight. The force field artillery headquarters provides the fires cell with operational control of all corps fires. To further facilitate fires when the corps is serving as the senior Army tactical headquarters, the air support operations center colocates with the fires cell. The fires cell provides representatives to the current operations integration cell. The fires cell includes the elements shown in figure 2-5 (page 2-17).

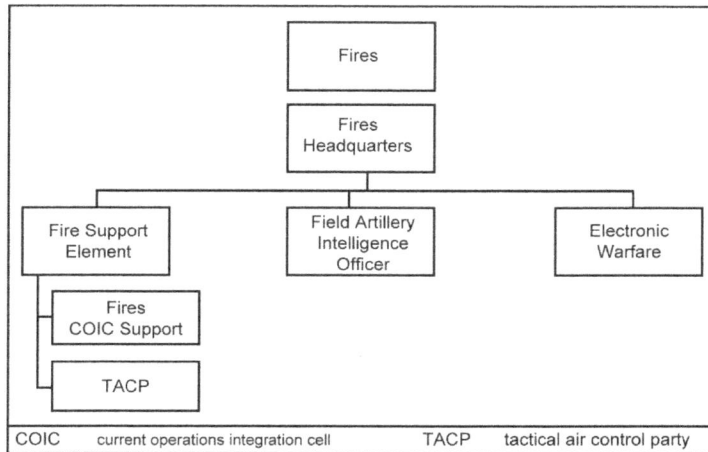

Figure 2-5. Main command post fires cell

Fires Support Element

2-77. The fires support element synchronizes military and civilian, joint, and multinational lethal fires and nonlethal actions and field artillery sensor management. It provides input to intelligence collection and the targeting process.

Fires Current Operations Integration Cell Support

2-78. This element from the fires cell provides the personnel to support the current operations integration cell to synchronize fires support for the corps.

Tactical Air Control Party

2-79. The tactical air control party at the main CP is the senior Air Force element in the corps and is organized as an air execution cell. It can request and execute Type 2 and Type 3 controls of close air support missions. (See JP 3-09.3.) Staffing is situation-dependent; however, the element includes, as a minimum, an air liaison officer and joint terminal attack controller. The element may also include Air Force weather and intelligence support personnel. (See appendix E for a greater discussion of the tactical air control party.)

Field Artillery Intelligence Officer

2-80. As a participant in the corps and joint targeting process, the field artillery intelligence officer coordinates with corps internal and external all-source intelligence elements. This officer provides input to the development, nomination, and prioritization of targets.

Electronic Warfare

2-81. The electronic warfare section supports the commander during full spectrum operations (offensive, defensive, and stability or civil support operations). Electronic warfare applies the capabilities to detect, deny, deceive, disrupt, or degrade and destroy enemy combat capabilities by controlling and protecting friendly use of the electromagnetic spectrum. These capabilities—when applied across the warfighting functions—enable commanders to address a broad set of targets related to electromagnetic spectrum so gaining and maintaining an advantage within the electromagnetic spectrum.

PROTECTION CELL

2-82. The protection functional cell contains sections concerned with preserving the force so the commander can apply maximum combat power. It coordinates, integrates, and monitors military and civilian, joint and multinational protection support for corps units and installations. The cell provides protection functional expertise and advises the CG in developing essential elements of friendly information, the defended asset list, and the critical asset list. The protection functional cell provides vulnerability mitigation measures to help reduce risks associated with a particular course of action and conducts planning and oversight for full spectrum operations. Representatives from the protection cell may provide input to plans and future operations cells, depending on the operational environment and the commander's preference. Commanders tailor and augment the protection cell with functional expertise to form a protection working group as the mission requires.

2-83. The protection functional cell coordinates with the command and control cell concerning information protection tasks. The cell coordinates with the surgeon concerning preventive medicine. The cell coordinates with the fires cell concerning integrated fires protection (formerly called counter-rocket, -artillery, and -mortar), including planning and coverage areas.

2-84. The protection cell includes the sections shown in figure 2-6.

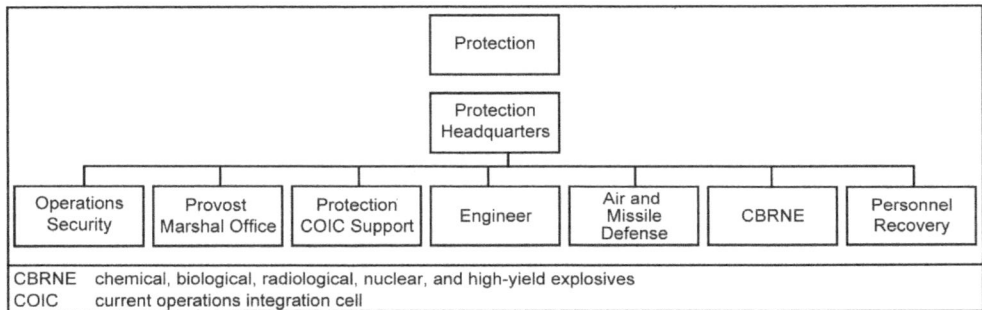

Figure 2-6. Main command post protection cell

2-85. The main CP protection functional cell performs these tasks in support of corps operations:

- Directs the coordination, planning, and analysis of protection activities.
- Monitors operational security activities, including identification of essential elements of friendly information.
- Coordinates chemical, biological, radiological, nuclear, and high-yield explosives (CBRNE) activities, including planning and information dissemination.
- Synchronizes and integrates military police activities, including detainee and enemy prisoner of war operations.
- Synchronizes and integrates engineer operations.
- Coordinates air and missile defense operations.
- Integrates personnel recovery operations into orders and plans.
- Develops and monitors safety programs for the command.

Operations Security

2-86. The operations security section—

- Coordinates operations security activities within the corps headquarters and CPs.
- Conducts vulnerability analysis, assesses the corps' operations security risks, and monitors implementation of operations security control measures by corps forces.

2-87. The operations security section includes a counterintelligence branch. This branch detects, identifies, tracks, exploits, and neutralizes the multidiscipline intelligence activities of friends, competitors, opponents, adversaries, and enemies. Focusing primarily on passive counterintelligence, it coordinates with the counterintelligence coordinating authority (in the intelligence cell) to deconflict actions.

Provost Marshal Office

2-88. The provost marshal is the principal advisor to the CG on military police functions. This officer plans, analyzes, coordinates, and monitors military police functions within corps forces. Military police functions include police intelligence, law and order, internment and resettlement, maneuver and mobility support, and area security operations. (See FM 3-39.) The corps provost marshal may serve as the chief of detainee operations when the corps is the senior Army unit in a joint operations area with detainees. (See JP 3-63.)

Current Operations Integration Cell Support

2-89. The protection current operations integration cell support section provides personnel who work in the current operations integration cell to synchronize protection operations for the corps.

Engineer Section

2-90. The engineer section in the protection cell integrates the engineer functions (combat, general, and geospatial engineering) within the protection function by—
- Coordinating and synchronizing engineer efforts in support of protection (such as survivability and environmental considerations) throughout the headquarters, between echelons, and with multinational forces and governmental and nongovernmental organizations.
- Advising the chief of protection on—
  - Construction requirements and standards associated with survivability efforts (such as hardening facilities).
  - Engineer capabilities available or needed to meet protection requirements.
  - Environmental considerations.

Air and Missile Defense

2-91. The air and missile defense section oversees corps air and missile defense operations. The section coordinates the four primary air and missile defense missions: air and missile defense, situational awareness, airspace management, and force protection. It coordinates these missions by—
- Coordinating air and missile defense activities with other CP cells, especially regarding airspace command and control and aviation operations.
- Disseminating weapons control status and the air tasking order.
- Coordinating with the area air defense commander on all land-based and air and missile defense within the corps AO.

For further information air and missile defense within a corps AO, see appendix D.

Chemical, Biological, Radiological, Nuclear, and High-Yield Explosives Section

2-92. The CBRNE section—
- Advises the commander and staff on CBRNE issues.
- Plans combating weapons of mass destruction elimination operations (coordinates for disposal of weapons of mass destruction).
- Provides oversight on weapons of mass destruction elimination operations.
- Plans for sensitive site assessments operations, tracks sensitive site exploitation operations, and provides reachback technical support.
- Performs information superiority analysis (tracks key indicators).

- Conducts CBRNE response analysis.
- Provides CBRNE defense, obscuration, and flame input to estimates, orders, and plans.
- Provides consequence management planning, support, and analysis.
- Plans support for joint operations.
- Provides support to Army organizations, as required.
- Identifies explosive ordnance disposal requirements and recommends and implements explosive ordnance disposal unique skills to protect the force.
- Tracks, prioritizes, and reinforces support to counter unexploded explosive ordnance, improvised explosive devices, and weapons of mass destruction.

Personnel Recovery Section

2-93. Personnel recovery is inherently a joint operation. The personnel recovery section coordinates corps personnel recovery activities with joint, multinational, and host-nation personnel recovery operations. (See JP 3-50 and FM 3-50.1.) Personnel recovery tasks include—

- Developing and maintaining the corps personnel recovery program, including procedures, planning, preparation, execution, and assessment.
- Coordinating personnel recovery issues with higher, lower, and adjacent organizations.
- Establishing a joint personnel recovery center, if directed.

Safety Element

2-94. Safety, although not a part of the corps table of organization and equipment, is an important augmentation to the corps staff. The safety officer is part of the commander's personal staff. A safety element works within the protection cell to assist the commander and staff with integrating composite risk management for training and operations. (See FM 5-19.)

SUSTAINMENT CELL

2-95. The sustainment functional cell contains sections that provide support and services to ensure the corps's freedom of action, extend its operational reach, and prolong its endurance. Four staff sections contribute sections to the sustainment cell: assistant chief of staff, personnel (G-1); assistant chief of staff, logistics (G-4); assistant chief of staff, financial management (G-8); and the surgeon. The G-4 serves as both the chief of sustainment cell and the logistic section chief. Elements perform specific functions within each staff sections shown in figure 2-7 (page 2-21). (See appendix A for a detailed discussion of corps sustainment.)

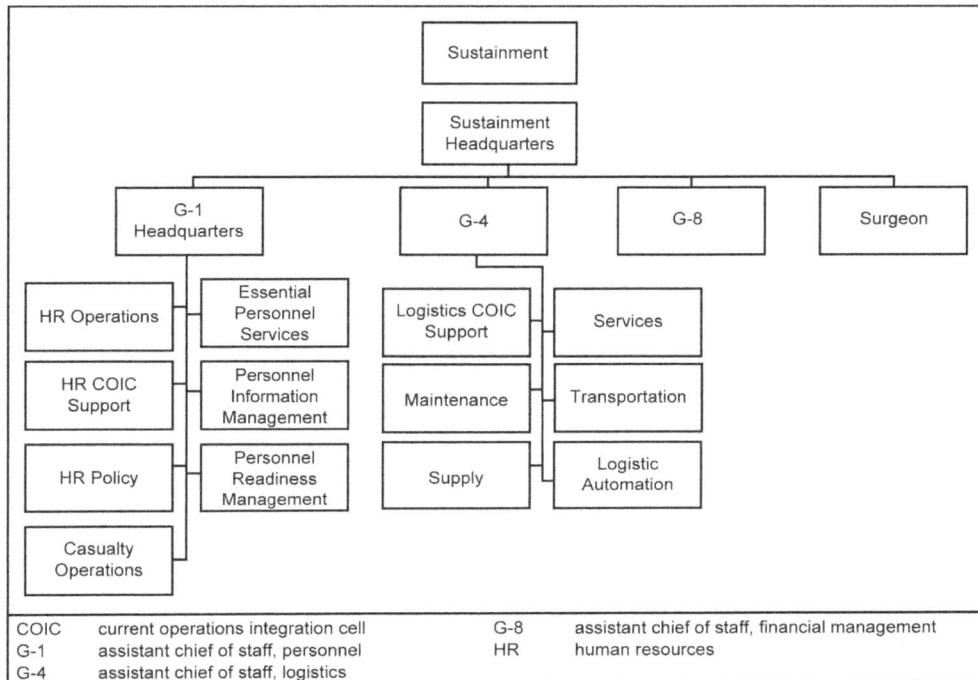

Figure 2-7. Main command post sustainment cell

2-96. The sustainment functional cell performs these tasks to support corps operations:

- Develop and implement human resource policies and procedures.
- Coordinate personnel support.
- Monitor the human resources situation and provide input to common operational picture (COP).
- Coordinate casualty operations.
- Synchronize and integrate logistics operations to include maintenance, supply and services, transportation, general engineering, and mortuary affairs.
- Provide logistics input to the COP.
- Coordinate resource and financial management operations.
- Synchronize and integrate Army health system operations.

Human Resources Section

2-97. The G-1 is the corps CG's principal human resources advisor and the chief of the human resources section. (See FM 1-0.) This section establishes human resources policies and ensures human resources support is properly planned, resourced, and executed for corps forces. In addition, the corps human resources section—

- Establishes human resources policy.
- Conducts essential personnel services.
- Coordinates morale, welfare, and recreation.
- Conducts casualty operations.
- Performs strength reporting and personnel readiness management.
- Conducts personnel information management.

- Manages headquarters manning.
- Receives and processes individual augmentees.
- Coordinates band operations.
- Performs command interest programs.
- Monitors postal operations.

Human Resources Operations Element

2-98. The human resources operations element conducts morale, welfare, and recreation operations; manages command interest programs; manages retention efforts; and monitors postal operations for corps units. It manages the information assurance program and civilian personnel programs for Department of Defense personnel.

Human Resources Current Operations Integration Cell Support

2-99. The G-1 current operations integration cell support element provides personnel who work in the current operations integration cell to synchronize human resources operations for the corps.

Human Resources Policy

2-100. Human resources policy responsibility involves developing, coordinating, and managing current, mid-range, and long-term human resources personnel policies for the corps. It includes providing oversight for executing human resources activities for corps units.

Casualty Operations

2-101. Casualty operations include collecting casualty information for preparing estimates, reporting casualties, and conducting notification and assistance programs. Casualty information is provided by casualty liaison teams, medical treatment facilities, mortuary affairs, and reports from corps forces.

Essential Personnel Services

2-102. The branch establishes, processes, and manages essential personnel services for the corps units. This responsibility includes establishing processing priorities and timelines for submitting actions by corps forces. The branch processes personnel actions requiring the CG's signature.

Personnel Information Management

2-103. Personnel information management involves collecting, processing, storing, displaying, and disseminating human resources information about corps Soldiers, units, and civilians. This function includes maintaining the human resources information systems. The human resources cell coordinates with the command and control cell as necessary to establish communication links.

Personnel Readiness Management

2-104. Personnel readiness management involves analyzing personnel strength data to determine current combat capabilities, projecting future requirements, and assessing conditions of unit individual readiness. Personnel readiness management directly interrelates and depends on the functions of personnel accountability, strength reporting, and personnel information management. Strength reporting reflects the combat power of a unit using numerical data.

Logistics Section

2-105. The G-4 oversees the corps logistic elements. This section oversees—

- Logistic operations.
- Maintenance.
- Supply and services.

- Transportation.
- Logistic automation.

See FM 4-0 for details on sustainment.

Logistics Current Operations Integration Cell Support

2-106.   The G-4 current operations integration cell support element provides personnel to work in the current operations integration cell. The support provided helps synchronize logistic operations for the corps.

Maintenance

2-107.   The logistic element performs the following maintenance-related functions:

- Formulating policy, procedures, and directives related to materiel readiness.
- Providing oversight of equipment and ordnance maintenance, recovery, and salvage operations.
- Participating in joint, inter-Service and host-nation agreements to provide resources to support corps operations.
- Monitoring and analyzing maintenance functions and equipment readiness status.

Supply and Services

2-108.   The logistic element performs the following supply and services-related functions:

- Formulating and implementing policy and procedures for the classes of supply (less class VIII) and related services.
- Monitoring corps logistic operations regarding—
  - Supply systems.
  - Transportation networks.
  - General engineering.
  - Maintenance.
  - Miscellaneous services (mortuary affairs, food service, billeting, textile repair, clothing exchange, and laundry and shower).

Transportation

2-109.   Transportation operations involve advising the corps CG on the following to support deployment and redeployment of forces and distribution of materiel:

- Transportation policy.
- Transportations systems.
- Movement planning and execution.
- In-transit visibility.
- Automated systems.

2-110.   The logistic element coordinates with internal and external entities regarding mobility operations. This element includes the Air Force air mobility liaison officer who advises the corps CG on airlift activities.

Logistic Automation

2-111.   The logistic element monitors and reports the status of corps logistic automated information systems.

Financial Management Section

2-112.   The G-8 is the corps CG's principal advisor on financial management and chief of the financial management section. This section obtains guidance on policy, appropriations, and funding levels and provides guidance to tactical financial managers. It estimates, tracks, and reports costs for specific operations to support requests to Congress as required. This element establishes the aggregate levels of fiscal support to be allocated and imposes directed resource constraints. It provides input to the program objective memorandum, prepares budget schedules and adjusts budgets based on program budget decisions. The corps G-8 chairs funding boards for corps forces. In addition, the G-8 is responsible for the following elements:

- Plans and operations.
- Budget execution.
- Special programs.

Surgeon

2-113.   The surgeon is charged with planning for and executing the Army Health System mission within the corps. (See appendix A for additional information.) The surgeon performs the following functions:

- Advises the corps commander on the health status of the command.
- Monitors, prioritizes, synchronizes, and assesses Army Health System support.
- Serves as medical contact officer for the corps.
- Provides an analysis of the health threat.

COMMAND AND CONTROL CELL

2-114.   Command and control of the corps is the duty of the CG. The CG is the focus of the command and control system and the supporting warfighting functions. See figure 2-8 (page 2-25). The command and control cell tasks for the corps include:

- Synchronize and integrate information engagement components (public affairs, combat camera, strategic communications, defense support to public diplomacy, and leader and Soldier engagement).
- Integrate civil affairs operations.
- Coordinate psychological operations.
- Coordinate network operations support.
- Plan and execute computer network defense.
- Synchronize and integrate information assurance.
- Establish and monitor information protection and communications security.

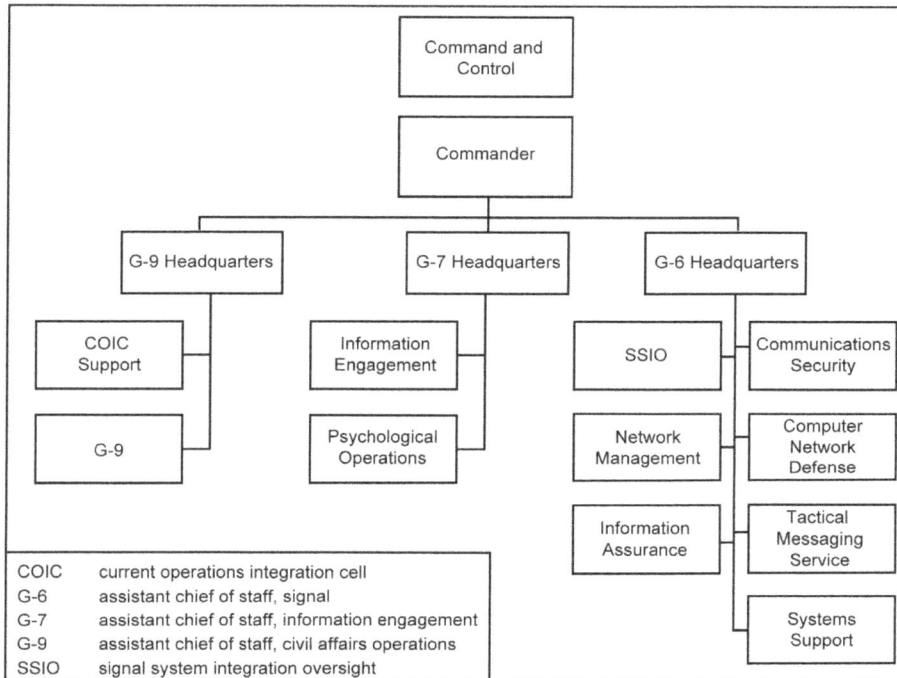

Figure 2-8. Main command post command and control cell

Civil Affairs

2-115. The assistant chief of staff, civil affairs operations (G-9) integrates civil affairs operations functions and capabilities into corps operations. The civil affairs operations staff serves in the command and control cell in the main CP under the chief of staff. The civil affairs operations staff provides representatives to the current operations integration cell as a current operations integration cell support element from civil affairs. Civil affairs operations staff members are also assigned in the future operations and plans cells in the main CP. Additionally, civil affairs operations staff members participate in meetings (including boards and working groups) as needed. The civil affairs operations staff performs these functions (see FM 3-05.40):

- Advises the corps commander on allocating and using civil affairs units under corps control.
- Develops the civil affairs operations annex to corps plans.
- Recommends civil affairs operations augmentation, including appropriate functional specialists.
- Reviews higher headquarters' plans.
- Informs the corps main CP staff on the civil affairs operations capabilities and units.
- Supports ISR activities of the corps.
- Shares enemy information and possible indicators and warnings collected through passive observation by the G-9 staff and supporting civil affairs operations units.
- Coordinates with the fires cell for lethal and nonlethal target development, measures of effectiveness, and synchronization of nonlethal activities with lethal fires, ensuring that civilian property, public buildings, and infrastructure are protected to the maximum extent possible.
- Coordinates and synchronizes corps civil affairs operations with higher headquarters' civil affairs efforts.

- Establishes a civil-military operations center or coordinates with an existing center to perform collaborative planning and coordination with interagency, intergovernmental, nongovernmental, and host-nation organizations.
- Analyzes how civilians impact military operations and how military operations impact civilians.
- Provides civil affairs analysis to meetings (including boards and working groups).
- Chairs the civil affairs operations working group, if formed.
- Supports the integrating cells.

Information Engagement

2-116. *Information engagement* is the integrated employment of public affairs to inform U.S. and friendly audiences; psychological operations, combat camera, U.S. Government strategic communication and defense support to public diplomacy, and other means necessary to influence foreign audiences; and, leader and Soldier engagements to support both efforts (FM 3-0). The assistant chief of staff, information engagement (G-7) oversees information engagement activities. (See FM 3-0.) The information engagement staff serves in the command and control cell in the main CP. As required, this staff augments the current operations integration, future operations, and plans cells. The staff also provides expertise to meetings (including boards and working groups).

2-117. The G-7 (assistant chief of staff, information engagement)—

- Serves as the principal information engagement advisor to the corps CG and staff principal advisors.
- Advises the corps CG on allocating and employing information engagement capabilities.
- Integrates information engagement into corps operations.
- Provides representation in the ISR operations section of the intelligence cell.
- Coordinates information engagement actions directly with adjacent and subordinate unit staffs.
- Chairs the corps information engagement working group.
- Participates in the targeting board and other boards and working groups, as required.

2-118. The psychological operations element conducts and assesses psychological operations capabilities within the main CP. The element leader advises the CG, coordinates with the G-7, and supervises the psychological operations element.

Signal

2-119. The assistant chief of staff, signal (G-6) oversees all corps communications, command and control information systems, and information management systems. The assistant chief of staff for signal is the chief of the command and control signal section. This section includes the elements shown in figure 2-10 (page 2-33).

2-120. The G-6 (assistant chief of staff, signal)—

- Advises the CG, staff, and subordinate commanders on communication and information networks.
- Directs development of network requirements and estimates.
- Oversees network planning.
- With the assistant chief of staff for operations and chief of the knowledge management element, establishes procedures for developing the COP.
- Oversees the management of corps internal networks.
- Coordinates external network support to the corps.
- Plans, manages, and executes electromagnetic spectrum operations.

Signal System Integration Oversight

2-121. The signal system integration oversight element involves the following:

- Providing technical staff support to Army and joint units allocated to the corps.
- Overseeing network certification and integration.
- Monitoring the state of network modernization, readiness, communications-electronics maintenance, and sustainment.
- Overseeing contractor support of the corps network.
- Coordinating commercialization of corps communication and information technology capabilities.
- Supervising the installation of corps main and tactical CPs wire and cable networks.

Network Management

2-122. The network management element involves the following:

- Managing the corps network, from the applications residing on corps platforms through the points at which the corps network connects to the Global Information Grid.
- Maintaining network connectivity to all corps forces, including deployed units, units en route to the theater of operations, and units at home station.
- Monitoring network performance and quality of service, including interoperability of the corps network with external networks not controlled by the corps.
- Managing frequency assignments for the corps.
- Deconflicting electromagnetic spectrum for all corps emitters.
- Supervising delivery of defense message system services to the main and tactical CPs.
- Coordinating with the knowledge management section to develop and align tactical network enforceable information dissemination management policies and services.

Information Assurance

2-123. The information assurance element ensures the availability, integrity, reliability, authentication, and nonrepudiation of information. It does the following:

- Coordinates command and control information systems interface with joint and multinational forces.
- Develops, promulgates, and monitors information assurance policies.
- Oversees performance of communications security functions.

Communications Security

2-124. The communications security element provides communications security operational support and facilitates communications security planning for corps forces. Associated responsibilities include the following:

- Receiving, transferring, accounting, safeguarding, and destroying communications security materials for the corps CPs.
- Providing training and instructions to communications security hand-receipt holders and users in the proper handling, control, storage, and disposition of communications security materials.
- Performing communications security key compromise recovery and reporting of communications security incidents.

Computer Network Defense

2-125. The computer network defense element—

- Establishes network defense policies (such as, accreditation, information assurance vulnerability assessment compliance, and access control).
- Provides staff oversight of the network defense policy implementation by corps forces.
- Ensures establishment and maintenance of security boundaries for network operations under corps control with military and civilian, joint and multinational organizations.
- Manages corps headquarters intrusion detection systems.

Tactical Messaging Service

2-126. The tactical messaging service element provides tactical defense message system services to the main and tactical CPs. Services include access to the defense message switch global address directory and the capability to send and receive signed and encrypted record message traffic.

Systems Support

2-127. Signal system support teams from the headquarters battalion signal company perform—

- Coordinate and supervise construction, installation, and recovery of cable and wire communications systems and auxiliary equipment within main and tactical CPs.
- Install and operate the corps information technology help desk.
- Provide voice, video teleconference, e-mail—Non-Secure Internet Protocol Router Network (NIPRNET), SECRET Internet Protocol Router Network (SIPRNET), and other communication networks—assistance, and other help desk functions.
- Assist Army and joint forces under corps control with network installation and troubleshooting as directed by the assistant chief of staff for signal.

MAIN COMMAND POST INTEGRATING CELLS

2-128. All cells and elements of the corps CPs are responsible for integrating information. The three integrating cells of the main CP apply information from the functional cells to the corps's operations with regards to time. As depicted in figure 2-9 (page 2-29) these cross-functional cells—current operations integration, future operations, and plans—coordinate across the artificial boundaries often created by functional designations.

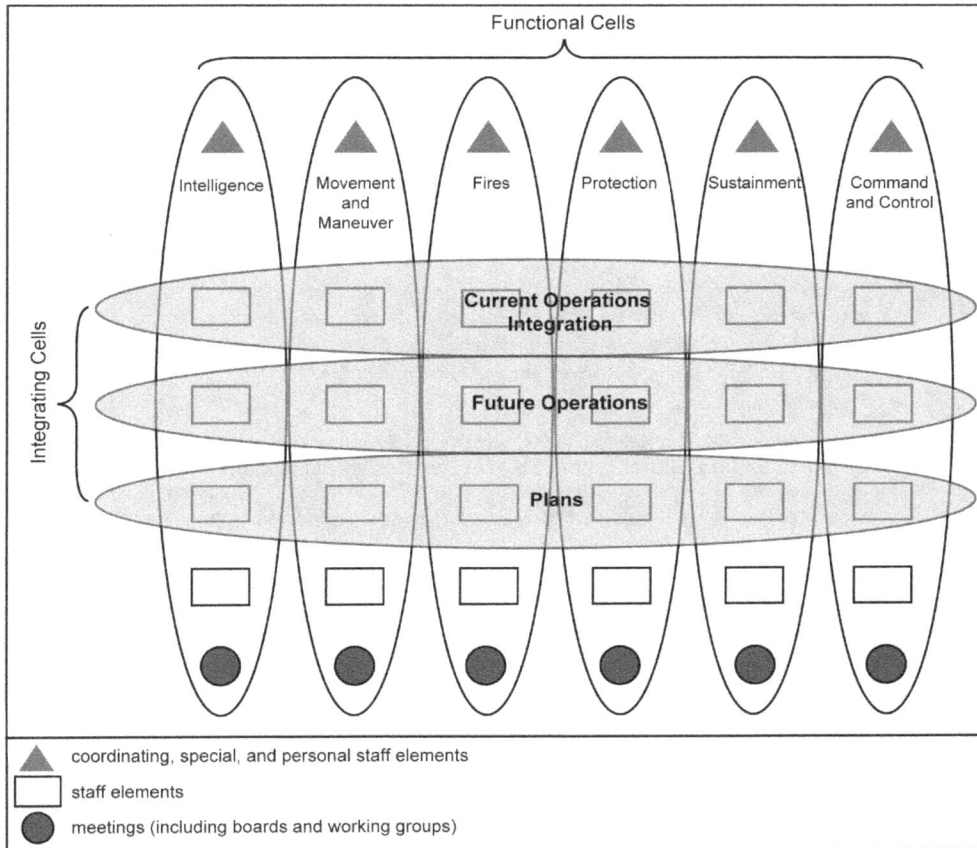

Figure 2-9. Main command post integrating and functional cells

CURRENT OPERATIONS INTEGRATION CELL

2-129. The current operations integration cell monitors the operational environment and directs and synchronizes operations in accordance with the concept of operations and commander's intent. Current operations focus on the "what is," and rapidly progress through the decision cycle through executing battle drills. Current operations produce a larger volume of orders including administrative fragmentary orders and tactical and operational fragmentary orders. The current operations integration cell is the element that synchronizes all staff actions. This cell is the hub of daily activities within the corps main CP. Members of functional, special, and personal staffs who provide their support to ongoing operations augment the current operations integration cell. The cell is led by the chief of operations (a colonel), a lieutenant colonel deputy, and an operations noncommissioned officer. The cell's planning horizon is hours and days. Primary tasks include the following:

- Implement and maintain the current operations cell standing operating procedures.
- Assess the tactical situation.
- Recommend the commander's critical information requirements and keep them current.
- Maintain the COP.
- Build and maintain the battle rhythm.

- Control operations by issuing operation and fragmentary orders.
- Serve as the central clearinghouse for incoming messages, orders, requests for information, and taskings.
- Synchronize actions among the other CP cells, elements, meetings (including boards and working groups), and other entities, such as personal and special staff sections that operate independently.
- Establish and conduct liaison operations to and from corps units and other organizations.
- Provide the current situation to the future operations and plans cells to support their operations.
- Perform near-term task assessment.

2-130. The current operations integration cell staffs these positions for 24-hour coverage:

- Operations officers.
- Battle command officers.
- Request for information and orders managers.
- Request for information and orders noncommissioned officers.
- Operations sergeants.

2-131. The noncommissioned officers in the current operations integration cell should be graduates of the battle staff noncommissioned officer course. The CG augments the current operations integration cell as necessary. Augmentees may include representatives from other main CP integrating cells, the other Services, other government agencies, and multinational partners to include the host-nation.

FUTURE OPERATIONS CELL

2-132. The future operations cell plans and assesses operations for the mid-range planning horizon. Future operations focus on the "what if" and normally move relatively slowly, with more deliberate assessment and planning activities. For a corps, this planning horizon is days and weeks. The cell's primary tasks include the following:

- Serving as the link between the current operations integration cell and plans cell.
- Monitoring current operations.
- Contributing to the COP.
- Turning command guidance into executable orders.
- Modifying plans to support current operations.
- Assisting in or producing fragmentary orders to support current operations.
- Developing branches to current operations.
- Recommending commander's critical information requirements.
- Conducting short- to mid-range planning to support current operations.
- Participating in the targeting process.
- Performing mid-term operations assessment.

2-133. The future operations cell includes a general plans element and a functional plans element. The cell is staffed with a mix of officers and noncommissioned officers led by a graduate from the School of Advanced Military Studies. When possible, the noncommissioned officers are graduates of the battle staff noncommissioned officer course. As with the current operations integration cell and plans cell, the future operations cell is augmented whenever a skill set—such as special operations—is required.

2-134. Normally ongoing operations mean that several plans undergo refinement simultaneously with associated working groups and joint planning teams. The work of these elements occasionally overlaps, and it is important that one individual or agency is appointed to maintain situational awareness of all planning efforts.

General Plans Element

2-135. The general plans element in the future plans cell includes a mix of majors led by a lieutenant colonel assisted by a senior noncommissioned officer. They develop and enhance the current order's common elements. This element provides the concept of operations that the functional plans element completes with detailed, specific knowledge.

Functional Plans Element

2-136. The functional plans element includes members who concentrate on the parts of planning that require functional expertise. Composed of majors with noncommissioned officer support, the functional plans element contains many specialties. These include intelligence, aviation, human resources, logistics, protection, command and control warfare, civil affairs operations, and special technical operations.

PLANS CELL

2-137. The plans cell is responsible for mid- to long-range planning. It focuses on the "what's next," and interacts with the higher headquarters planning efforts. Using the military decisionmaking process, it develops complete operation plans, branches, and sequels. The cell monitors the COP for situational awareness. It stays abreast of current operations by coordinating with the current operations integration cell. It is staffed by specialists, including a planner who graduated from the School of Advanced Military Studies, a strategic plans officer, and an officer certified in the Joint Operation Planning and Execution System. Individuals with specific expertise from elsewhere in the main CP and from entities external to the corps augment the plans cell as required. The plans cell's primary tasks include the following:

- Monitoring current operations.
- Conducting tactical planning in support of major operations and battles.
- Conducting operational-level planning, to include developing supporting plans to higher headquarters plans.
- Coordinating with the current operations integration and future operations cells to understand the current situation and planned short-term activities.
- Supervising and coordinating the preparations for all operation plans and some branches.
- Managing Joint Operation Planning and Execution System planning, including input and review of the time-phased force and deployment data.
- Using the joint operations planning process to support joint requirements, activities, and processes. (See JP 5-0 for a discussion of the joint operations planning process.)
- Coordinating and managing force structure.
- Planning force management for corps forces to include Army Reserve and Army National Guard capabilities.
- Coordinating with respective theater army plans activities on all aspects of planning within their respective combatant commander's areas of responsibility.
- Conduct military deception planning.

2-138. The assistant chief of staff, plans (G-5) leads the plans cell. The cell has two elements: the plans element and the force integration element.

2-139. The deputy plans officer, a lieutenant colonel, leads the plans element. It is staffed by specialists in the following fields:

- Sustainment.
- Intelligence.
- Military deception.
- Civil affairs operations.
- Space and special technical operations.
- Information engagement.
- Psychological operations.

- Fires.
- Protection.
- Engineer.
- Aviation (augment).

2-140. The force integration element deals with the functions and processes employed in raising, provisioning, sustaining, maintaining, training, and resourcing corps forces in garrison and when deployed.

SECTION III – TACTICAL COMMAND POST

2-141. The corps tactical CP is organized as an additional current operations cell encompassing six functional cells. It is capable of 24-hour operations. This CP can control corps operations for a limited time and can form the nucleus of an early-entry CP. It normally colocates with the main CP but remains fully operational as a current operations cell so that it can operate independently. This requires staff, equipment, and procedures to be in place and fully exercised. When employed, the deputy assistant chief of staff for operations usually oversees its activities. The cell and section chiefs for the tactical CP are normally deputy coordinating staff officers, deputy CP cell chiefs, or officers with special expertise.

2-142. Figure 2-10 (on page 2-33) depicts a basic corps tactical CP. This CP may be established and task-organized for specific missions of varying duration with augmentation from planning, sustainment, civil affairs, or mobility staff elements. The headquarters battalion provides the tactical CP with a task-organized communications and life support element. The headquarters battalion commander designates a leader for the support element when deploying the tactical CP.

2-143. The tactical CP can provide command and control continuity while the main CP is moving. It can also be used—

- As a task force headquarters for a specific corps operation—such as a river crossing, vertical envelopment, or sensitive site exploitation—while the main CP retains control of the overall operation.
- In support of separate operations in a noncontiguous AO.
- In control of a shaping or sustainment operation while the main CP controls the decisive operation.

2-144. The tactical CP includes representatives from all six functional cells. Depending on the situation and with suitable augmentation, the tactical CP performs the same functions as the main CP.

INTELLIGENCE CELL

2-145. There are five elements in the tactical CP intelligence cell:

- Headquarters section.
- Target development.
- Fusion.
- Distributed tactical exploitation system.
- Staff weather office.

HEADQUARTERS

2-146. The headquarters element is led by the deputy assistant chief of staff for intelligence and oversees the intelligence cell. Its tasks include—

- Providing intelligence support for current operations.
- Interfacing with the movement and maneuver cell to integrate intelligence products and intelligence synchronization into current operations.

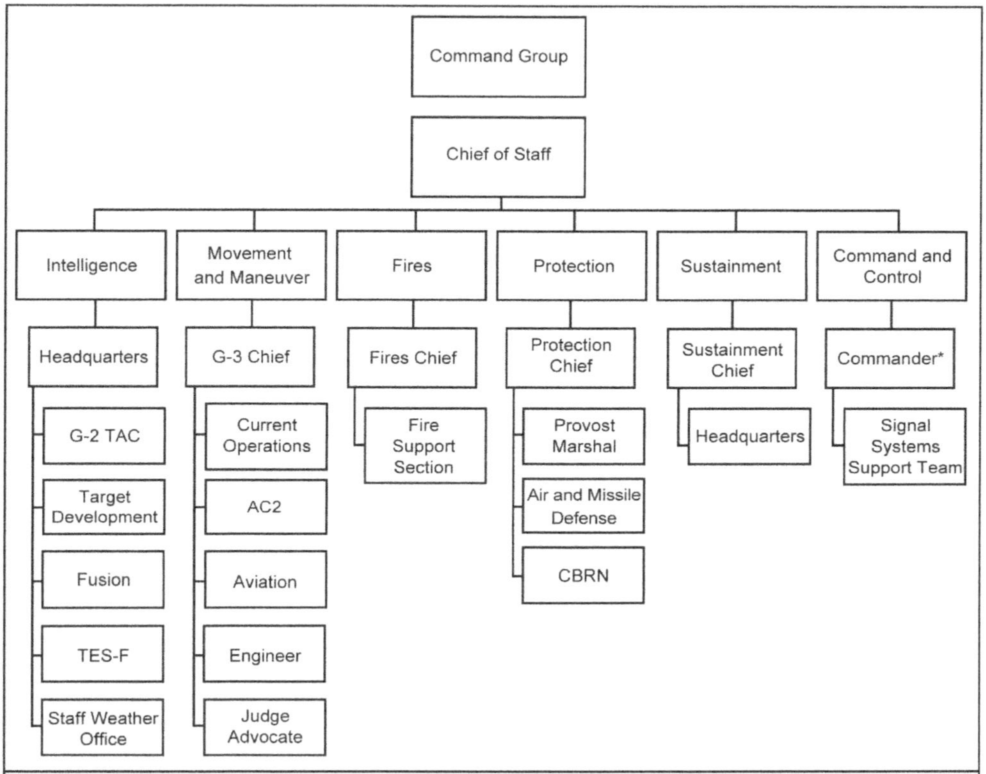

Figure 2-10. Corps tactical command post

TARGET DEVELOPMENT ELEMENT

2-147. The target development element—
- Participates in the targeting process.
- Develops and nominates priority targets.
- Integrates prioritized and sequenced targets into current operations.

FUSION ELEMENT

2-148. The fusion element receives, processes, analyzes, and disseminates all-source intelligence to support current operations.

TACTICAL EXPLOITATION SYSTEM–FORWARD

2-149. Using the Distributed Command Ground Station–Army, the distributed tactical exploitation system-forward receives, processes, analyzes, exploits, and disseminates the following products to support current operations:

- Signals intelligence.
- Imagery intelligence.
- Measurement and signature intelligence.
- Geospatial intelligence.

STAFF WEATHER OFFICE

2-150. The staff weather office provides weather analysis and coordinates with other weather teams in support of the tactical CP.

MOVEMENT AND MANEUVER CELL

2-151. The tactical CP movement and maneuver cell consists of five elements:

- Current operations.
- Airspace command and control.
- Aviation.
- Engineer.
- Judge advocate.

CURRENT OPERATIONS ELEMENT

2-152. The current operations element oversees operations of the tactical CP. This includes the following:

- Monitoring ongoing corps operations.
- Directing current operations.
- Collecting relevant information.
- Producing and disseminating a COP when operating independently.
- Maintaining liaison with higher headquarters, subordinate units, and joint and multinational headquarters as appropriate.
- Synchronizing actions among the other CP cells, elements, meetings (including boards and working groups), and other entities (such as personal and special staff sections that operate independently).

AIRSPACE COMMAND AND CONTROL ELEMENT

2-153. The airspace command and control element oversees the airspace command and control function. This includes providing tactical airspace requirements to the main CP airspace command and control element and integrating tactical operations with the main CP airspace command and control element.

AVIATION ELEMENT

2-154. The aviation element coordinates and synchronizes the execution of aviation maneuver, unmanned assets, and sustainment operations when the tactical CP operates independently.

ENGINEER ELEMENT

2-155. The engineer element coordinates with the primary engineer element at the main CP and advises the CG and staff on current operations requirements to enable freedom of movement and maneuver and protection. This element also recommends how to allocate and use engineering assets to fulfill those near-term requirements. It coordinates the application of the engineer functions (combat, general, and geospatial engineering). With a focus on the movement and maneuver function, the engineer element coordinates and synchronizes engineer operations within the tactical CP and with other headquarters, between echelons, and with multinational forces and governmental and nongovernmental organizations as required.

JUDGE ADVOCATE ELEMENT

2-156. The judge advocate element provides support in all legal disciplines, including advice on—
* Rules of engagement.
* Rules of law activities.
* Stability operations.
* Treatment of detainees.
* Lawfulness of targets and weapons.
* The law of war.

FIRES CELL

2-157. The tactical CP fires cell is led by the deputy chief of fires who may locate elsewhere as the situation requires. The cell consists of an assistant fire support coordinator, fire support noncommissioned officers, and fire support specialists. Normally, the tactical CP fires cell executes lethal fires and nonlethal activities for a specific operation or for short durations. The fires cell may require additional augmentation from the main CP fires cell, depending on mission requirements.

2-158. The fires cell—
* Plans fires.
* Requests and coordinates close air support and air interdiction.
* Interfaces with the battlefield coordination detachment, air liaison officer, Army and joint airspace control elements, and higher and joint fires elements.
* Coordinates air requirements through the battlefield coordination detachment at the ARFOR liaison to the joint force air component command with the corps air liaison officer.
* Synchronizes Army, joint, interagency, and multinational lethal fires and nonlethal activities.
* Conducts combat assessments and recommends reattacks.
* Coordinates with special operations forces in the corps AO.
* Coordinates with Army and joint airspace control elements.
* Coordinates with higher and joint fires elements.

PROTECTION CELL

2-159. The protection cell includes the following elements:
* Provost marshal office.
* Air and missile defense.
* Chemical, biological, radiological, and nuclear (CBRN).

PROVOST MARSHAL ELEMENT

2-160. The provost marshal element does the following:
* Tracks current operations.
* Recommends the employment of military police assets.
* Coordinates military police support for ongoing operations, including processing of detainees.

AIR AND MISSILE DEFENSE ELEMENT

2-161. The air and missile defense element plans, coordinates, and synchronizes air and missile defense operations. This includes the following:

- High-value target protection analysis.
- Coordination with main CP and external organizations.

CHEMICAL, BIOLOGICAL, RADIOLOGICAL, AND NUCLEAR ELEMENT

2-162. The CBRN element is responsible for the following:

- Providing 24-hour operations.
- Advising the CG and staff on all CBRN issues.
- Overseeing immediate CBRN logistics functions.
- Contributing to information superiority analysis-track key indicators.
- Performing CBRN response analysis.
- Coordinating for and providing input to estimates (intelligence and vulnerability analysis), orders, and plans.
- Recommending the employment of CBRN assets.
- Supporting the CBRN warning and reporting system managed by the main CP.

SUSTAINMENT CELL

2-163. The sustainment cell monitors, collects, assesses, prioritizes, and executes all sustainment functions at the tactical CP, including movement operations, asset in-transit visibility, and requirements estimation. It provides input to the COP.

COMMAND AND CONTROL CELL

2-164. The command and control cell includes support from the corps headquarters battalion tactical CP signal systems support team. As with the main CP command and control cell, the tactical CP can include civil affairs, psychological operations, and information engagement elements if required. The command and control cell—

- Synchronizes and integrates information engagement components (public affairs, combat camera, strategic communications, defense support to public diplomacy, and leader and Soldier engagement).
- Integrates psychological operations.
- Integrates civil affairs operations.
- Coordinates network operations support.
- Plans and executes computer network defense.
- Synchronizes and integrates information assurance.
- Establishes and monitors information protection and communications security.
- Installs, operates, maintains, and defends NIPRNET and SIPRNET. Manages the installation and operation of local area networks, including cable and wire installation and troubleshooting.
- Monitors, manages, and controls organic communications systems.
- Performs, as required, network operations functions, including enterprise management, electromagnetic spectrum operations, information dissemination management, and information assurance.
- Installs, operates, and maintains information services within the information network enabling all battle command functions across all corps and below formations.

SECTION IV – ADDITIONAL COMMAND AND CONTROL FACILITIES

2-165. There are two temporary command and control facilities that may see action depending on the situation: the mobile command group and the early-entry CP. Subordinate forces may form a third type of facility, the center, to exercise command and control of all or part of a function.

MOBILE COMMAND GROUP

2-166. The mobile command group is the CG's mobile CP. It permits the CG to exercise command and control while on the move. The mobile command group has two armored wheeled vehicles, each with a crew and elements of the Army Battle Command System suite of terminals. Its composition depends on the CG's desires and the situation. The group is augmented as required, including additional vehicles, communications equipment, and security. If aerial transportation is required, the corps requests aviation assets from the theater aviation command or brigade or other aviation assets. (See figure 2-11 for a possible mobile command group composition.)

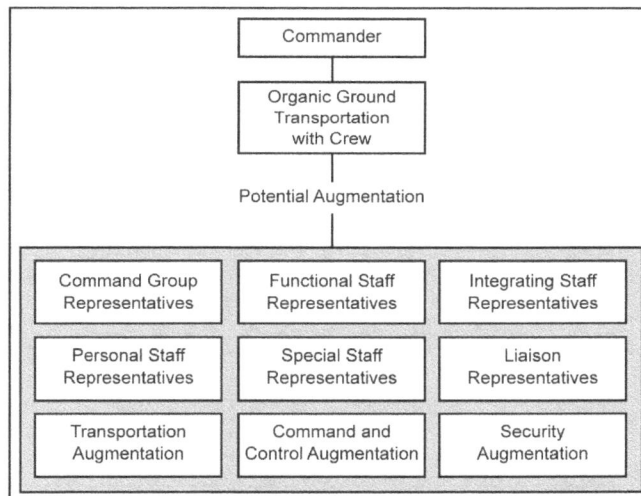

Figure 2-11. Mobile command group

EARLY-ENTRY COMMAND POST

2-167. Occasionally the CG needs to establish a CP at a remote location ahead of other corps headquarters elements. In those circumstances, the corps creates and deploys an early-entry CP. The early-entry CP may deploy and set up alone. Alternatively, it may colocate with another Army CP, a joint CP, an interagency or multinational organization, or a host-nation command and control facility.

2-168. The early-entry CP is usually organized and tailored around the tactical CP. It draws equipment and personnel from the tactical CP, main CP, and other corps communications and security elements. The CG staffs the early-entry CP with a mix of current operations personnel able to coordinate the reception of the corps and plan its initial operations. The corps standing operating procedures normally designate the sources of personnel and equipment. Depending on the situation, the CG may augment the early-entry CP with such capabilities as language and regional expertise.

2-169. The corps early-entry CP performs the functions of the main and tactical CPs until those CPs are deployed and fully operational. A deputy commander, chief of staff, or assistant chief of staff for operations leads the early-entry CP.

CENTERS

2-170. A center is a command and control facility established for a specific purpose. Centers are similar to CPs in that they are facilities with staff members, equipment, and a leadership component. However, centers have a more narrow focus and are normally formed around a subordinate unit headquarters. For example, a civil affairs unit under corps control may establish a civil-military operations center or a logistics organization supporting the corps may establish movement control centers.

SECTION V – AIR FORCE SUPPORT TO THE CORPS

2-171. Air Force support to the corps headquarters consists of a corps tactical air control party, staff weather officer, and the air mobility liaison officer. These Air Force elements function as a single entity in planning, coordinating, deconflicting, and integrating the air support operations with ground elements. Air mobility liaison officers advise ground commanders and staffs on the capabilities and limitations of air mobility assets. The Air Force provides tactical control parties to Army maneuver unit headquarters down to the combined arms battalion. Additionally, the Air Force pools terminal attack control teams to provide support down to the maneuver company level as required. See appendixes B and E for additional information. When serving as a senior tactical echelon the corps will normally have an air support operations center.

SECTION VI – CORPS HEADQUARTERS AND HEADQUARTERS BATTALION

2-172. The only troops organic to the corps are in the headquarters battalion. The battalion is commanded by a lieutenant colonel. It consists of the battalion command group, headquarters battalion staff, and four companies. (See figure 2-12.) The subordinate elements of each company report to the company chain of command. The company commander, in turn, reports to the headquarters battalion commander.

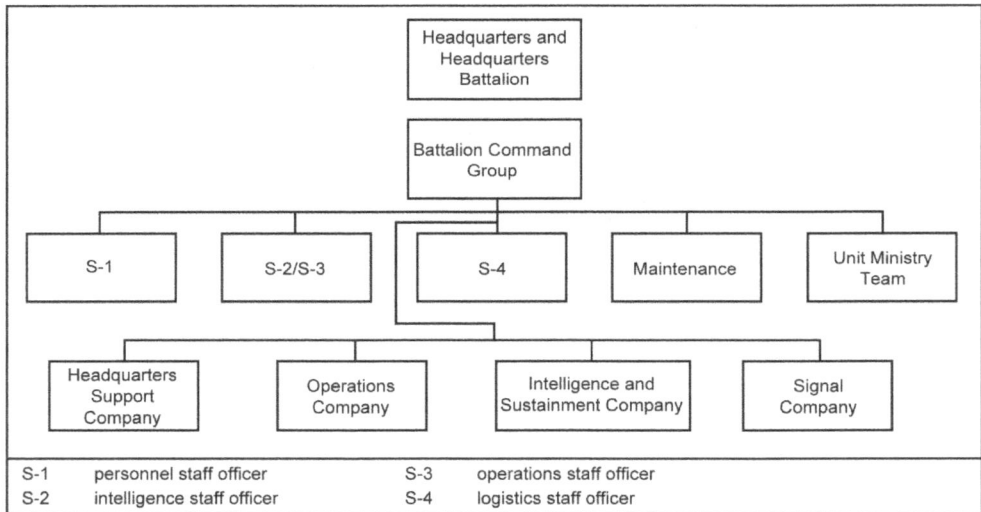

Figure 2-12. Headquarters battalion

BATTALION COMMAND GROUP

2-173. The battalion command group provides supervision and exercises command and control of personnel assigned to the corps headquarters. It consists of a commander, executive officer, command sergeants major, rear detachment commander, rear detachment noncommissioned officer, and a vehicle drive. The battalion commander also serves as the headquarters commandant for the corps headquarters.

2-174. The battalion provides communications, transportation, and medical support to the corps headquarters. The battalion's personnel, equipment, and services are split among the main CP, tactical CP, and mobile command group. The battalion provides administrative and life support to the additional resources assigned or attached to the corps headquarters—such as a band, security assets, and joint or interagency augmentation—as required.

BATTALION STAFF

2-175. The headquarters battalion staff consists of five staff sections. They provide administrative, human resources, logistic support, religious support, and life support to corps headquarters elements in garrison and the field. When deployed, the battalion staff sections are responsible for unit-level command and control, communications support, property accountability, transportation, medical, food service, and maintenance support for the main CP, tactical CP, and mobile command group.

HEADQUARTERS SUPPORT COMPANY

2-176. The headquarters support company contains the components shown in figure 2-13. All members of the headquarters battalion staff are assigned to the headquarters support company.

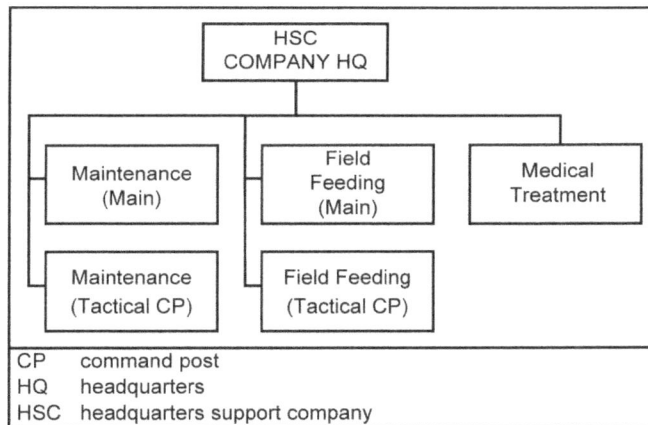

Figure 2-13. Headquarters support company

MAINTENANCE SECTIONS FOR MAIN AND TACTICAL COMMAND POSTS

2-177. The headquarters support company has two sections that support headquarters battalion Soldiers in the corps CPs, one for the main CP and one for the tactical CP. When deployed these sections provide command and control and the following support to the personnel assigned to the respective CPs: maintenance and field feeding.

MEDICAL TREATMENT SECTION

2-178. The medical treatment section provides Army Health System support for the corps main CP personnel as well as emergency and advanced trauma management to main and tactical CP personnel. It also provides sick call services, medical surveillance and preventive medicine, and unit-level ground and en route patient care.

OPERATIONS COMPANY

2-179. The operations company (A Company) provides company-level administrative and logistic support to Soldiers in the movement and maneuver cell, protection cell, and fires cell, as well as the tactical CP elements of t hese cells. When the tactical CP de ploys, t he headquarters a nd he adquarters battery commander may direct the operations company headquarters to deploy as well, to synch ronize all asp ects of support to the tactical CP from the headquarters and headquarters battery. (See figure 2-14.)

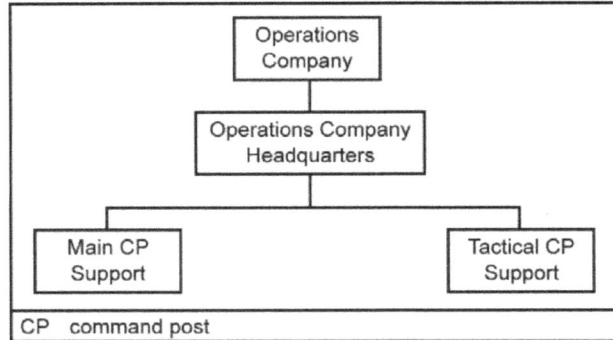

Figure 2-14. Operations company

INTELLIGENCE AND SUSTAINMENT COMPANY

2-180.   The in telligence an d su stainment co mpany (B Co mpany) p rovides co mpany-level admin istrative and logistical support to the Soldiers in the intelligence cell and the sustainment cell, as well as the tactical CP ele ments of t hese cells. When t he tactical CP deploys, t he headquarters a nd h eadquarters battery commander m ay d irect the in telligence and sustain ment headquarters to d eploy as well so to syn chronize all aspects of support to the tactical CP from the headquarters and headquarters battery. (See figure 2-15.)

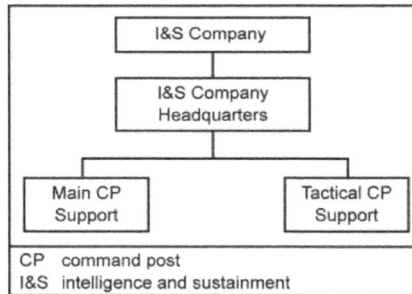

Figure 2-15. Intelligence and sustainment company

SIGNAL COMPANY

2-181.   The si gnal c ompany pro vides i nformation net work a nd com munications su pport t o t he c orps headquarters. It includes platoons that di rectly support the cor ps main and t actical C Ps. (See fi gure 2-16, page 2-41.) Si gnal Sol diers supporting t he corps CPs are assigned t o and receive their administrative support from the signal company.

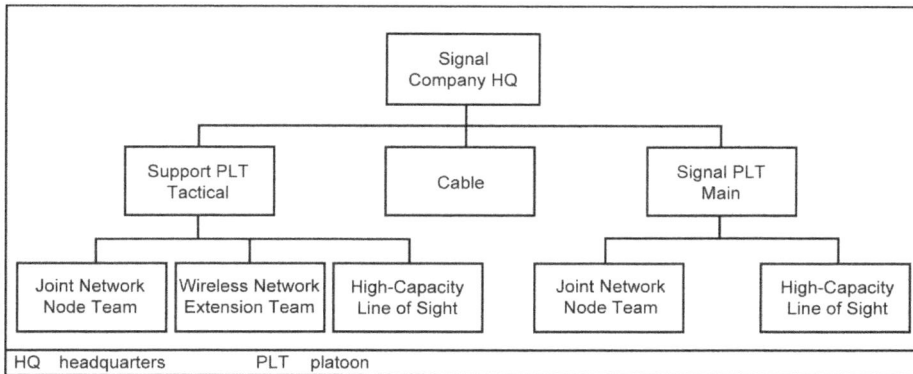

Figure 2-16. Signal company

COMPANY HEADQUARTERS

2-182.   The signal company headquarters provides logistic support to the company. The company receives maintenance support from the headquarters battalion. The company headquarters oversees installation and operation of the following support for the main and tactical CPs, and others as directed:

- Network.
- Radio (line-of-sight and satellite communications).
- Wireless network extension.
- Wire.
- Cable.

MAIN AND TACTICAL COMMAND POST SUPPORT PLATOONS

2-183.   Each CP support platoon provides communications support using the joint network node to connect user devices such as telephones and computers. The two platoons have nearly identical capabilities to provide terrestrial and space-based communications support to each CP. Platoon capabilities include secure tactical defense switched network voice, NIPRNET, SIPRNET, Joint Worldwide Intelligence Communications System, and video teleconferencing. The tactical CP has the wireless network extension teams for extended frequency modulation retrains. The platoon cable section provides support to the main CP and tactical CP on a mission basis.

CIVILIAN, CONTRACTOR, AND OTHER AUGMENTATION

2-184.   The Army is supported by Army civilians, contractors, and other partners during peace and war. As combat multipliers, they perform critical duties in virtually every facet of Army operations at home and when deployed.

ARMY CIVILIANS

2-185.   Army civilians have long been an integral part of the Army team. They fill key leadership and support positions. The augmentation for a typical corps includes civilian employees in almost every part of the organization. This is true especially for a corps based overseas. There, many civilians serve in host-nation relations, maneuver management, programs such as Partnership for Peace, and plans and operations. Many of these positions require continuity, such as historian, protocol, safety office, training division, special security office, budget office, and transportation.

2-186. Army civilians with critical skills may deploy. The corps headquarters plans for civilian deployment regardless of the projected or actual corps mission. The headquarters identifies emergency essential civilians during deployment planning. Emergency essential positions meet two criteria. They cannot be converted to uniformed positions without a loss of continuity of performance. And they are required to ensure the success of combat operations or to support combat-essential systems. In unforeseen situations, Army civilians in positions not previously identified as essential can be required to deploy.

2-187. During military operations and once medically cleared, Army civilians fall under the military chain of command. They perform their specialty just as they would before deployment with regards to evaluations, assignments, discipline, and recognition. The corps is responsible for predeployment training, life, and other support for deployed Army civilians, including physical security. (See Department of the Army Pamphlet (DA Pam) 690-47 for more information.)

CONTRACTORS

2-188. Contractors' contributions to deployed forces include support other than direct participation in hostile actions in sustainment, language services, communications, and infrastructure.

2-189. Frequently, a desire to limit the presence of U.S. forces in a region leads to a cap on the number of Service members deployed. When military force caps are imposed, contractor support allows commanders to maximize the number of combat Soldiers by replacing military support units with contractor support. This force-multiplier effect lets the combatant commander provide sufficient support in the theater of operations while strengthening the deployed force's fighting capability. At the conclusion of operations, contractors also facilitate early redeployment of forces.

2-190. In the initial stages of an operation, supplies and services provided by local contractors improve response time and free strategic airlift and sealift for other priorities. Contractor support drawn from resources in the theater of operations augments existing support capabilities to provide a new source of critically needed supplies and services.

2-191. The corps should identify and provide qualified personnel as required by supporting contracting units and organizations to serve as contracting officer representatives. These representatives must be familiar in the goods or services provided by the contracts they help administer as well as trained, appointed, and managed by a contracting officer to whom they report. Contracting officer representatives ensure contractors provide the goods and services as specified in the supported unit's statement of work, thereby ensuring the supported unit receives the support necessary to support their mission.

UNIFIED ACTION

2-192. Within the U.S. Government, Army and other government agencies perform in both supported and supporting roles with other commands and agencies. However, this support command relationship differs from that described in joint doctrine. Relationships between the Army and other government agencies and organizations do not equal the command and control of a military operation. Whether supported or supporting, close coordination between the military and other agencies is key. (See JP 3-08.)

2-193. Coordination and integration among the joint force and other civil and military, joint and multinational organizations does not equal the command and control of a military operation. Military operations depend upon a command structure that differs from that of civilian organizations. These differences may present significant challenges to coordination efforts. The various government agencies' different—and sometimes conflicting—goals, policies, procedures, and decisionmaking techniques make unity of effort a challenge. Still more difficult, some intergovernmental and nongovernmental organizations may have policies that are explicitly antithetical to those of the United States, and particularly U.S. forces.

Chapter 3

Corps Headquarters Operations

This chapter describes the corps headquarters operat ions as a command and control headquarters. It highlights the im portance of the su pport to the corps headqu arters and their command and support relationship. It discusses the placement of the corps headquarters for command and control during operations and also how the corps uses its command and control systems throughout the operation.

SUPPORT TO CORPS HEADQUARTERS

3-1. Exercising command and c ontrol of land forces fo r operations is t he c orps hea dquarters' first priority. The corps is capable of carrying out this mission alone, but it is often augmented with personnel to assist the commander a nd staff in accomplishing their assigned tasks and missions. The types and qua ntity of au gmentation depends on t he si tuation an d normally arrives from theater a rmy assets or from th e available Army force pool. The augmentation timeframe can be temporary or can last the entire operation. For example, the plans cell and the movement element of the movement and maneuver cell may receive the greatest command attention and resources during the predeployment phase. The augmentation to the corps headquarters may be l imited i n t ime and sc ope t o s upport t he s uccessful de ployment of t he corps headquarters. While de ployed an d en gaged i n su stained major c ombat l and o perations i n an are a of operations (AO), the c urrent ope rations integration ce ll may be the ce nter of attenti on and receive the majority of augmentation

3-2. In its role as an intermediate tactical headquarters, the corps headquarters can exercise command and control of mixed brigades and divisions, as well as joint or multinational forces supported by theater assets. To c ontrol thi s force m ix, the corps he adquarters se rves as a hiera rchical orga nization c ombining the commanding general (C G), staff cells, a nd associated liaison elem ents in to an in tegrated who le. As such the corps headquarters organizes to control a wide array of assets. To command and control this force mix, the commander relies on staff execution and optimizing available command and control systems within the corps headquarters. Staff execution through functional and integrating cells, to include personal and special staff coordination, enables the commander to control a wide array of assets.

3-3. Corps headquarters a ugmentation ha ppens d uring al l p hase o f predeployment, depl oyment, and redeployment. D uring m ajor operations. Assets not av ailable i n t heater are requested f rom t he force generating bas e and f rom A rmy co mmands and direct r eporting u nits (See Fi eld M anual (FM) 1 -01). Representative force generating force organizations include—

- Headquarters, Department of the Army.
- Headquarters, United States Army Forces Command.
- United States Army Training and Doctrine Command.
- United States Army Reserve Command.
- United States Army Special Operations Command.
- United States Army Materiel Command.
- United States Army Medical Command.
- United States Army Space a nd Missile Defense Command/United States Army Forces Strategic Command.
- United States Army Network Enterprise Technology Command/9th Signal Command (Army).
- United States Army Intelligence and Security Command.
- Military Surface Deployment and Distribution Command.

- Headquarters, United States Army Criminal Investigation Command.
- United States Army Installation Management Command.
- United States Army Cadet Command.
- United States Army Corps of Engineers.

3-4. Available brigade and lower echelon units are generally part of theater-level commands tailored into smaller components to meet the corps's mission requirements. For example, the requirements for chemical, biological, radiological, nuclear, a nd hi gh-yield expl osives de fense m ay not require an e ntire che mical brigade, only a small component such as a biological hazard detection team.

3-5. In its turn, the corps headquarters task-organizes the divisions, brigade combat teams, and functional and su pport brigades pr ovided to it b y t he D epartment of the Arm y, the theate r a rmy, and other force generators for employment in land operations in an AO.

COMMAND AND SUPPORT RELATIONSHIPS

3-6. The command relationship of units providing support to the corps headquarters varies depending on the situ ation. Forces av ailable to th e corps h eadquarters fo r lan d operations are assig ned, attach ed, or placed under corps ope rational or t actical control. (See FM 3-0 for a di scussion of command and s upport relationships.)

3-7. The Army d evelops th e cap ability to rapid ly tailo r and task -organize ex peditionary forces (see FM 3-0). Eac h expe ditionary force i s a fl exible, m odular or ganization. The t heater a rmy i s t he Arm y Service com ponent c ommand of t he geographic com batant com mand. I n t hat r ole, t he t heater arm y exercises administrative control over all Army forces in the combatant commander's area of responsibility. The theater army tailors available fo rces to support corps headqua rters. Each corps ca n control a mix of divisions, brigade c ombat t eams, and functional an d s upport bri gades as an i ntermediate land force headquarters or a joi nt task force. Fi gure 3 -1 (page 3 -3) portrays the command and su pport rel ationships, including operational co ntrol (OPC ON), associated with a corps headquarters serv ing as an in termediate tactical headquarters.

3-8. The corp s h eadquarters normall y exercises command and control of land forces at the briga de a nd division levels. Its maneuver, sustainment, and other forces are normally attached, OPCON, in support, and occasionally under tactical control to the corps headquarters.

COMMAND AND CONTROL

3-9. Exercising c ommand and c ontrol i s a dy namic proces s i n whi ch t he corps st aff su pports t he C G throughout the operations process. The speed and accuracy with which the staff plans, prepares, executes, and assesses contributes to the CG's situational understanding. Commanders use several processes to solve problems: design, the military decisionmaking process, rapid decisionmaking and synchronization process, and Army probl em sol ving. Ad ditionally, t he C G an d st aff i ntegrates com posite risk m anagement throughout this process. FM 5-0 discusses these processes at length.

3-10. The corps relies o n inform ation, kno wledge, and b attle co mmand syste ms an d staff activ ities to support the execution and assessment of operations. In full spectrum operations, the corps has three goals for ex ecution: it ex tends th e operational reach o f its forces, sy nchronizes ope rations, and prioritizes and allocates resources. T he CG and c orps staff balance th ese g oals on ly with ad equate in formation. They receive information from several knowle dge management and information system s. Through i nformation management, the corps headquarters provides relevant information to the right person at the right time in a usable form to facilitate situational understanding and decisionmaking.

3-11. Effective co mmand and co ntrol of un its attach ed, under O PCON, under tactical contr ol, and in support to the corps headquarters also relies on effectively placing the corps main command post (CP). Its placement m ust m aximize the full capabilities of the corps com mand and control syste m and communications supp ort syste m. Th e co rps g ains ex ecution supp ort fro m the Arm y Battle Co mmand System, th e comman d po st of th e fu ture, battle rh ythm, o perations synchronization m eetings, and battle update briefings.

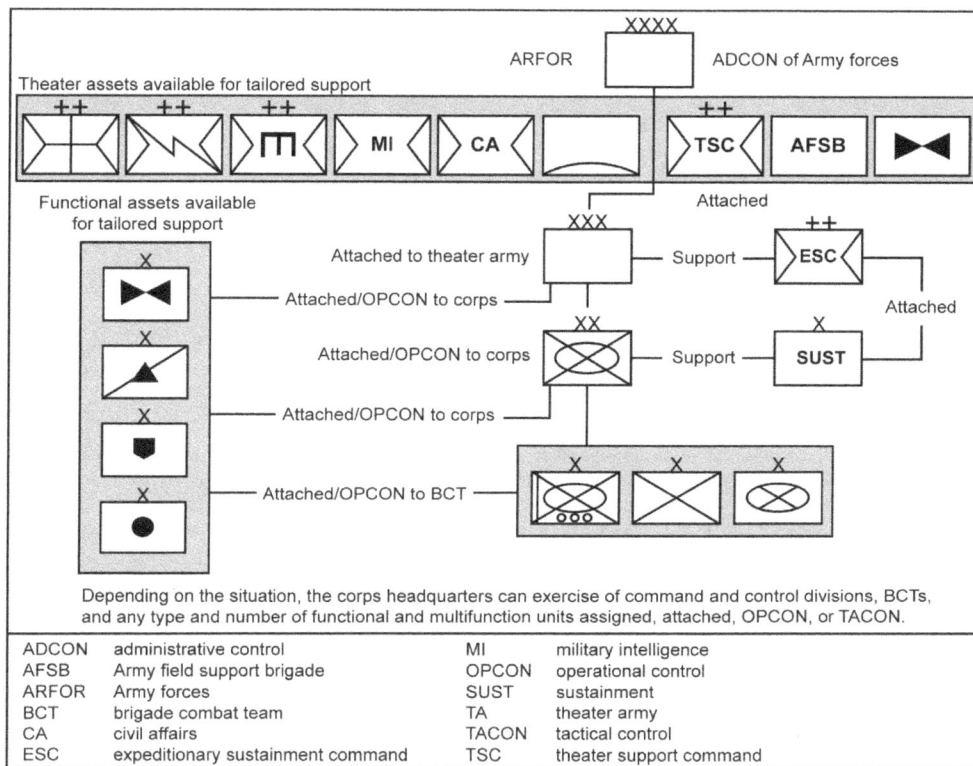

Figure 3-1. Corps command and support relationships

COMMAND POST PLACEMENT

3-12. The corps' placement of its main CP is vital to the success of command and control. The location of the main CP depends on such factors as communication reliability, security, concealment, and accessibility. The CG's location depends on the situation. A routine mission on the less violent end of the spectrum of conflict, such as a joint or multinational training exercise, may place less of a demand on the CG for rapid decisions and the necessary situational understanding. Conversely, a major combat operation against a near-peer enemy may demand frequent guidance and ever-changing decisions. The level of trust the CG has in the staff and their level of training influences this decision. Well-trusted senior leaders on a well-trained staff allow the CG more freedom of action.

3-13. The location of the corps main CP also affects how the CG commands the operation. The corps CP can be located in any of a number of places, including—

- As a joint headquarters colocated with the geographic combatant command.
- Colocated with the theater army CP.
- Located in a sanctuary or staging base remote from the AO.
- Located within the corps AO with noncontiguous subordinate command AOs.
- Positioned within the corps AO with contiguous subordinate units.
- Colocated with a division headquarters or other subordinate CP in a contiguous AO.
- Colocated with a subordinate CP in an AO with noncontiguous subordinate unit AOs.

3-14. Each CP location has its own distinct communications, travel, security, and time requirements. Secure satellite communications can reduce the impact of distance from the CG's location to that of superior or subordinate commanders, but it cannot eliminate the distance requirement. Travel from a sanctuary or a corps main CP location separated by hours from that of its subordinates adds to the time the commander is away from the main CP. Travel also adds to integrating and synchronizing the operation, reviewing plans and orders, and contributing to non-routine decisionmaking. In addition, the CG does not travel alone, and the staff, no matter how small, requires long-haul vehicles, aircraft, and en route security precautions that may make it inconvenient to travel long distances, especially if the enemy situation is uncertain. The CG and senior leaders and staffs weigh the benefit gained by command presence visits by the CG at remote subordinate CP locations with the loss of communication, the time the CG is away from the activities at the main CP, and the chance of capture or worse.

3-15. Regardless of the location selected by the CG, there must be a plan for continuity of command and control. Command and control continuity has two requirements. The first is to have a properly designated commander available to command, including a predesignated succession of command. The second is to organize the command and control system so the CG can exercise that authority continuously. Continuity depends on alternate and redundant facilities, time for transitions, and mitigating the effects of sleep deprivation.

3-16. The corps tactical CP is not designed to replace the main CP for extended periods, but it must be trained and ready to assume control of operations for short periods or special concurrent missions. Usually the tactical CP is colocated with the main CP and is operational at all times. The tactical CP becomes the corps command and control hub when the main CP is unavailable or when the corps forms a separate command and control entity for a specific operation. The latter might be running an air or sea port of debarkation or a distinct mission such as humanitarian assistance. As discussed in earlier chapters, the corps tactical CP is organized as a current operations cell capable of 24-hours-a-day operations. Unless employed for an extended period, the tactical CP does not deal with planning or the transitions from plans to operations.

3-17. The corps tactical CP maintains the same level of situational awareness as the current operations integrating cell and other elements of the main CP. To do so, the tactical CP establishes a transfer standing operating procedure. This standing operating procedure ensures the tactical CP replicates all the information systems that support command and control. Each of the six tactical CP cells coordinates with its main CP counterpart. Depending on the current situation, capabilities not programmed for the tactical CP but required to conduct current operations may deploy with the tactical CP to support the corps mission. The CG aims to continue to lead the corps and the staff to retain the capacity to control it. Maintaining the common operational picture (COP) and other situational awareness activities at the main CP during the time it is not in control reduces the time the main CP requires to achieve full readiness and effectively reassume command and control responsibilities.

3-18. Transferring and maintaining information is crucial during any handover of control between the main and the tactical CPs. The corps automated information systems allow information to be entered from several points at once into a central processing system. This electronic collaboration simplifies the handover of information and provides a level of information assurance to protect all information systems when the tactical CP controls operations. Each tactical CP functional cell ensures that the digital and other information systems databases appropriate to their area of expertise are maintained while the tactical CP is controlling operations.

3-19. Dynamic and fast-paced operations require the corps staff to keep abreast of operations. Depending on the time the tactical CP is expected to control operations in the absence of a fully operational main CP, portions of the plans and future operations cells may deploy with the tactical CP to sustain continuity of the planning effort. Cell members may deploy as individuals to become a part of the tactical CP movement and maneuver cell or form a provisional plans cell in the tactical CP. When the tactical CP controls operations, the headquarters battalion elements that support it are tailored to support augmentation from the plans cell, liaison sections, and other capabilities that accompany the CP to the new location. As with digitization, sustainment and security are integrated into the separated deployment of the tactical CP.

COMMAND AND CONTROL SYSTEMS

3-20. Command and control systems (see paragraph 2-19 for the definition of command and control systems) enable the commander to arrange personnel, information management, procedures, equipment, and facilities essential for the commander to conduct operations. Because of the complexity of commanding and controlling a mixed force conducting full spectrum operations, success of the corps headquarters relies on commanders effectively using available command and control systems. By using the Army Battle Command System, command post of the future system, and other communications support to the corps headquarters, the commander, staff, and subordinate commands can share a COP during the conduct of operations.

Army Battle Command System

3-21. The Army Battle Command System (ABCS) is a collection of information management systems that provides automated network information systems for the corps and its subordinates to support the operations process. The goal of this system of systems is to facilitate Army operations by integrating information from internal and external sources. The ABCS provides a COP using a common map set and database to present a visual display of the AO. Additionally, the ABCS is a suite of common service tools. These collaboration tools include:

- Video teleconferencing.
- Interactive whiteboard.
- File transfer services.
- Calendar and schedule applications.
- Task management tools.
- Internet browsers.
- Database query tools.

3-22. When deployed, travel limitations, distance, physical security concerns, and other factors frequently limit face-to-face contact throughout the corps' AO and beyond. Therefore, collaboration must occur virtually through interactive tools available to the corps headquarters, such as the ABCS. The ABCS, along with the command post of the future, facilitates near real-time collaboration and enable effective battle command. They allow Army commanders at all echelons to provide a COP to higher and subordinate echelon commanders and their staffs. However, face-to-face is the preferred method and it is often required to build commander-to-commander personal relationships.

3-23. Figure 3-2 (page 3-6) diagrams the ABCS. These ten battlefield automated systems make up the capabilities required to support corps operations. The ABCS integrates the information systems that support the Army warfighting functions and link them to strategic, operational, and tactical headquarters.

3-24. The ABCS requires trained and skilled operators. The corps uses the system often in garrison functions, so that the transition to deployed operations causes minimal friction in the corps headquarters' ability to effectively exercise command and control. See FM 6-0 for a description of these systems.

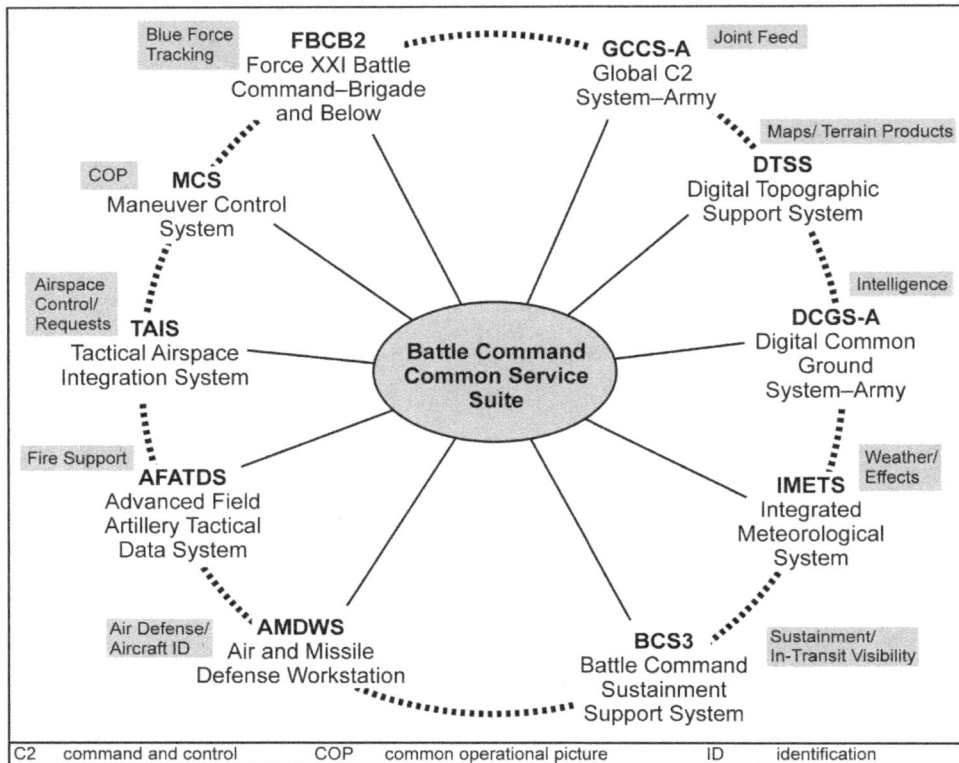

Figure 3-2. Army battle command system components

Command Post of the Future

3-25. The command post of the future is an executive-level decision support system that provides situational awareness and situational understanding for the commander and staff. This collaborative system is commonly used and provides additional capability for the commander to exercise command and control 24/7 throughout the AO. Key capabilities are—

- Second dimension and third dimension information visualization.
- Information liquidity-drag and drop information analysis across visualization products.
- Visibility of evolving understanding among distributed subordinates and team members.

3-26. Commanders often use the command post of the future to conduct battle updates, track enemy and friendly actions, and interface with the ABCS to allow sharing a COP. Efficient use of the system relies on trained and skilled operators. Commanders should use the command post of the future regularly in the garrison environment to ensure operators maintain skill proficiency.

Communications Support to Command and Control

3-27. An effective command and control system provides the CG with relevant information to adjust operations rapidly in response to changing situations. It informs staff members of the status of the operation, so they can communicate that information internally and externally to all echelons. This timely flow of information ensures all levels share a common understanding of the situation. The communications support to the corps' command and control system plays an important part in the equation. For commanders to exercise command and control throughout the operation, they need reliable and survivable communications.

3-28. Effectively integrating communication networks with information systems that support command and control enables the corps to manage, disseminate, and protect information throughout its AO. For example, LandWarNet, the Army's portion of the Global Information Grid, supports commanders by linking information to decisions and decisions to actions. This maturing capability unifies function-unique networks, interdependent battle command and information systems, and key network services to enable commanders, staffs, and subordinate units to collect, process, store, retrieve, disseminate, and protect information. LandWarNet connects all components and echelons of the generating force and the operational Army, giving corps operating in a joint environment access to global information resources and support services.

3-29. The CG and staff have access to a suite of communications systems as a part of the joint network node-network. The network provides the corps headquarters with a high-speed and high-capacity backbone of voice, video, and data communications tools designed to meet corps, division, and brigade battle command and information requirements. Communication links are provided by several satellite systems— Ku (Kurtz-under band) and Ka (Kurtz-above band) terminals, standard tactical entry point terminal, and extremely high-frequency band terminals. High-capacity line-of-sight communications systems are also available for use where appropriate. Figure 3-3 identifies standard corps systems (see FM 6-02.43 for details):

- Warfighter information network–tactical, increment one (known as WIN–T Inc 1).
- Satellite transportable terminal.
- High-capacity line of sight system.
- Secure, mobile, anti-jam, reliable, tactical terminal (known as SMART–T).
- Wideband satellite terminals.

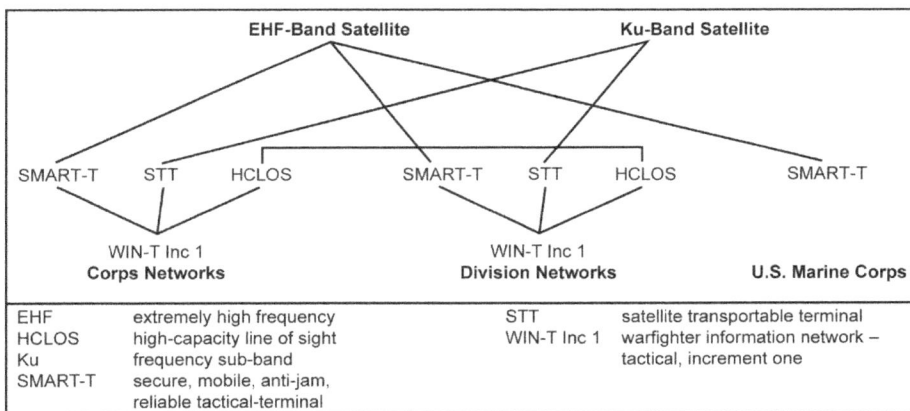

Figure 3-3. Corps communications

3-30. Warfighter information network–tactical, increment one refers to the communication equipment that provides switching for voice, video, and data communications. It also provides information assurance equipment to secure data and provides interoperability with legacy systems such as mobile subscriber equipment. It is a satellite communication and switching package that enables the corps, division, and brigade CPs to operate independently within the Global Information Grid or directly with a joint headquarters. Warfighter information network–tactical, increment one works with military or commercial satellites and ground systems.

3-31. The satellite transportable terminal uses a satellite transportable trailer equipped with a 2.4 meter Ku band satellite dish, which will be upgraded to Ka band as it becomes available. The network will work with existing terrestrial transport (high-capacity line-of-sight and line-of-sight) and satellite communications systems such as ground mobile forces (AN/TSC-85/93), tropospheric scatter (AN/TRC-170), and secure, mobile, anti-jam, reliable tactical terminal (AN/TSC-154).

3-32. The high-capacity line-of-sight system is a terrestrial microwave radio system paired with joint network node to provide high bandwidth line of sight capability.

3-33. The secure, mobile, anti-jam, reliant tactical terminal is a tactical military strategic satellite communication terminal which provides a satellite interface to permit protected uninterrupted voice and data communication as forces move beyond the line-of-sight capability of terrestrial communications systems fielded to corps, division, and brigade signal companies.

3-34. Wideband satellite terminals consist of older tactical satellite systems (AN/TSC-85D and AN/TSC-93D) used by theater-level expeditionary signal battalions to support corps and below operations.

3-35. LandWarNet subsystems directly impact the corps ability to communicate throughout the AO. Efficient and effective execution is supported by the voice, text, and imagery components that provide commanders and staffs accurate and timely information. Relevant information passed by all three modes of communication aims to be concise yet complete and clear enough to preclude misunderstanding. A COP designed to display information in an easy-to-understand format enables rapid adjustments during execution to keep an accurate portrayal of the situation in the AO.

BATTLE RHYTHM

3-36. The corps headquarters' battle rhythm supports the commander's effective command and control. Different battle rhythm may occur within the corps headquarters to accomplish simultaneous activities, but they all support the commander's overall battle rhythm. For example, the corps normally conducts a daily synchronization meeting and a battle update briefing to the commander or a designated officer to share the COP and receive guidance. The battle rhythm is both a process and the various forums identified as part of the process. When the battle rhythms of all the headquarters in the chain of command are nested with one another, they work more effectively. For example, unit activities are scheduled so that the information output in one activity is available as an input to higher or lower headquarters. Battle rhythms nested by echelon let corps subordinate commands offset their events to provide information needed by the corps headquarters. In addition, nested battle rhythms enable the corps CPs to supply necessary information to higher headquarters.

3-37. The battle rhythm is frequently portrayed on a daily basis, but it can be illustrated over weeks or months. An effective battle rhythm helps the CG and staff to synchronize the various information management processes—among them, update briefings, shift changes, and conference calls. A corps battle rhythm provides anchor points around which the CG and staff can plan their day. Chronological by nature, battle rhythm depends on inputs from earlier events and provides outputs needed for later events. Individual elements within the battle rhythm may be progressive; for example, a daily meeting, followed by a working group every three days, leading up to a board meeting every sixth day.

3-38. The battle rhythm operates best when the officer in charge can direct changes to fit the situation. As a guide to time management, battle rhythm is flexible enough to accommodate changes in the type of operation, availability of key individuals, and other interruptions in the routine. Failing to adhere to a disciplined battle rhythm results in the CG and staff working harder, longer, and less effectively. Establishing and maintaining a battle rhythm provides a disciplining mechanism to support rest and sleep plans for the CG and staff. This practice has the added benefit of training the second- and third-level leadership in the conduct of CP operations when the principals are not present.

3-39. A corps headquarters may have several types of battle rhythms: a live assembly of key individuals at a central location, a virtual meeting through use of the COP and the ABCS, or a combination of the two when the command group, primary, and selected others meet in a central location while others participate by video teleconference, Web camera, or ABCS. A battle rhythm is not a rigid tool to rob the CG of the opportunity to seize the initiative. The CG and staff need time to think. With the advent of ABCS and other command and control tools, the commander can receive the current COP at any time in any of the command and control facilities. Battle rhythm can be graphically depicted in a table, line, or circle. Figure 3-4 (page 3-9) illustrates a tabular battle rhythm. (See FM 5-0 for a further discussion of battle rhythm.)

Time	Event	Chair
0700	Battle update briefing	CG / DCG
0730	Shift change briefing / shift change in accordance with CP cell standing operating procedures	chief of staff / assistant chief of staff, operations
0800	Operations synchronization meeting	chief of staff/assistant chief of staff, operations
0900	Teleconference with higher headquarters	CG / DCG / chief of staff
	Common operational picture synchronization drill/virtual CP huddle	assistant chief of staff, operations
0930	Plans update	assistant chief of staff, plans
1000	Working groups/boards (on call of working group officer in charge)	officers in charge
1100	Running estimate updates	cell officers in charge
1300	Common operational picture synchronization drill / virtual CP huddle	assistant chief of staff, operations
1400	Working groups (on call of working group officer in charge)	officers in charge
1700	Common operational picture synchronization drill / virtual CP huddle	assistant chief of staff, operations
1900	Battle update briefing	CG / DCG
1930	Shift change briefing / shift change as per cell standing operating procedures	chief of staff / assistant chief of staff, operations
2100	Teleconference with higher (on call)	commander / deputy / chief of staff
	Common operational picture synchronization drill/virtual CP huddle	assistant chief of staff, operations
2130	Running estimate updates	cell officers in charge
2200	Working groups (on call of working group officer in charge)	officers in charge
0100	Common operational picture synchronization drill/virtual TOC huddle	assistant chief of staff, operations
0500	Running estimate updates	cell officers in charge
0600	Common operational picture synchronization drill/virtual CP huddle	assistant chief of staff, operations

CG	commanding general	DCG	deputy commanding general
CP	command post	TOC	tactical operations center

Figure 3-4. Example of corps battle rhythm

3-40. The battle rhythm can also be depicted in a circular fashion with internal activities inside the circle and external events on the outside. Figure 3-5 (page 3-10) shows an example of a circular or graphic battle rhythm chart. It assumes that the updated information disseminated at those briefings will enable the higher headquarters to receive the latest information when the corps CG communicates with the higher commanders and staff.

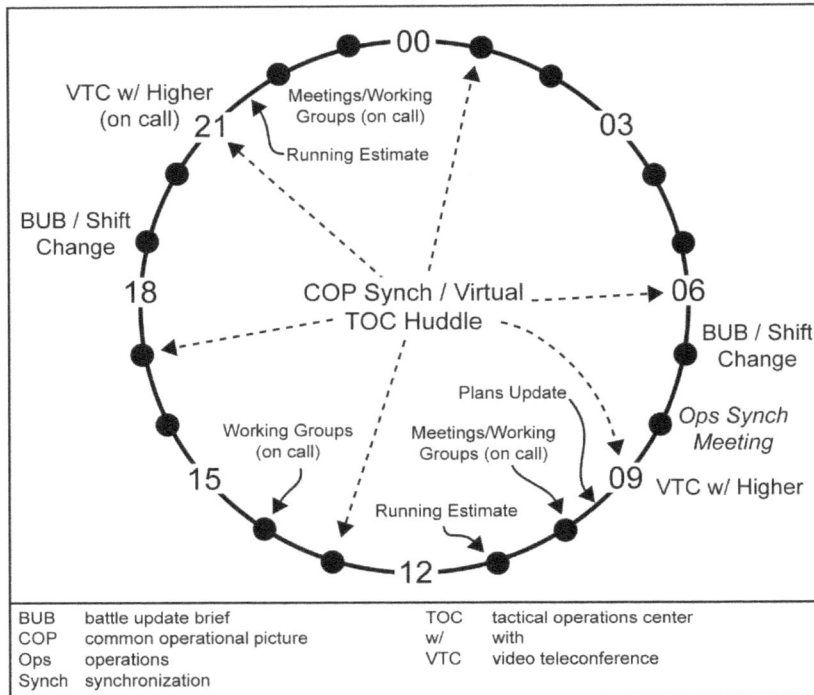

Figure 3-5. Example of a graphic battle rhythm

OPERATIONS SYNCHRONIZATION MEETING

3-41. Chaired by the corps chief of staff or assistant chief of staff for operations, the operations synchronization meeting is attended by senior leaders and is a key event in the corps headquarters' battle rhythm. The meeting provides information on the current operation. Principal members of the functional and integrating cells, separate staff sections, standing working groups and boards attend. They brief the status of ongoing projects and tasks with a focus on CG's priorities. The meeting's purpose is to synchronize warfighting functions for the short-term planning horizon and provide guidance that drives the operations of other components of the battle rhythm.

BATTLE UPDATE BRIEFING

3-42. The corps headquarters daily battle update briefing is an integral part of the corps headquarters ability to command and control. Because of the modern digital communications systems and databases available, the CG and staff can receive a battle update at any time. The CG is briefed in person, over a voice communications system, or by visual display. Typically part of the corps main CP battle rhythm, the battle update briefing provides analyzed information so the commander can make decisions and synchronize the staff's actions. Based on the updated COP, this briefing is intended to be short, informative, and selective. It provides the CG with limited information that addresses the current operation and activities planned for the near future. The corps tactical standing operating procedures, command guidance, and operational requirements determine what information is briefed. Normally, the CG reviews the status charts and displays before the battle update briefing to get familiar with the current situation of the corps. This enables the battle update briefing to focus on by-exception issues requiring CG attention and guidance.

Chapter 4

The Corps in Full Spectrum Operations

Often the corps is the highest echelon that commands and controls Army, joint, and multinational forces in offense or defense against an enemy or in a situation requiring stability or civil support operations. Meeting these challenges requires a corps headquarters able to assign missions to its subordinate formations, extend its operational reach, synchronize actions, and apply the elements of combat power.

EMPLOYING THE CORPS

4-1. Field Manual (FM) 3-0 articulates five operational themes: peacetime military engagement, limited intervention, peace operations, irregular warfare, and major combat operations. The themes are so interrelated that four of the five themes can occur at the same time. These themes give the commanding general (CG) and staff a way to characterize the dominant major operation underway in the corps area of operations (AO). These overlapping themes each require a different weighting of the elements of full spectrum operations. Often the main effort in each theme differs; the CG and staff shift the effort as the situation requires.

4-2. The Army's operational concept is full spectrum operations. Full spectrum operations span from benign, internationally sanctioned weapons inspections to major combat operations. Effective corps commanders exercise command and control for any operation. They must be as adept at planning a short noncombatant evacuation operation as supporting a multiyear major combat operation.

4-3. Corps headquarters controls Army and, when directed, joint and multinational forces and organizations when conducting (planning, preparing, executing, and assessing) full spectrum operations. The corps uses mission command to direct the application of full spectrum operations to seize, retain, and exploit the initiative through combinations of its four elements—offense, defense, and stability or civil support—and associated doctrinal tasks.

4-4. The corps headquarters is an essential element in the Army's expeditionary capabilities. These capabilities enable the Army to deploy combined arms forces into any operational environment and operate effectively upon arrival. Expeditionary operations require the corps and its subordinate forces to deploy quickly and shape conditions to seize the initiative and accomplish the mission.

4-5. Inherent in military operations is the reality of constrained resources. In most cases, too few human, material, and financial resources or too little time exists to support all operations equally. Units in the operational environment pool their resources to accomplish the mission rather than work independently. Corps forces are no exception. All corps operations aim to fully engage corps subordinate and supporting units in mission accomplishment.

4-6. As with many doctrinal tenets, operational reach, synchronization, and resourcing overlap. For example, the operational reach of the corps depends on factors such as the forces assigned, their ability to work together, and the priority of resourcing from the theater army or the joint headquarters.

EXTENDING OPERATIONAL REACH

4-7. *Operational reach* is the distance and duration across which a unit can successfully employ military capabilities (Joint Publication (JP) 3-0). Consistent with the higher commander's intent and mission, the CG tries to extend the corps' operational reach. The limit of the corps' operational reach is its culminating point—that moment in time and space when the force cannot continue its present operation. Operational reach depends on factors such as space, time, and available support. Corps operations design the operation so that its operational reach equals its culminating point. The corps expands the operational reach through the AO, available forces, a reserve, stability operations, forcible entry, and consequence management.

AREA OF OPERATIONS

4-8. A key component of operational reach is the corps AO and its associated *area of influence*—a geographical area wherein a commander is directly capable of influencing operations by maneuver or fire support systems normally under the commander's command or control (JP 3-16). The CG oversees and authorizes terrain management. This requires balancing the forces available with the size of the AO. Too large an area with regard to the available forces and the corps may not accomplish its mission. Too small an area and the corps will fail to use available forces as intended. One of the CG's first command decisions determines how to control the AO. Commanders also focus on their area of influence.

4-9. The CG begins terrain management by assigning subordinate forces to AOs. Such assignments empower individual initiative and maximize the opportunity for decentralized execution. When assigned an AO, the subordinate division or brigade commander takes responsibility for managing terrain, collecting intelligence, conducting security operations, tracking air and ground movement, clearing fires, and conducting operations in that AO. Normally, the corps also establishes a corps support area for the conduct of corps sustaining and enabling operations. (See FM 3-0 for a discussion of AO.)

4-10. The CG may divide the corps AO in one of three ways: contiguous AOs, noncontiguous AOs, or a combination of the two. Contiguous AOs enable subordinate units to share a common boundary. Noncontiguous AOs lack a common boundary between subordinate units. Combined contiguous and noncontiguous AOs contain some unassigned areas. (See figure 4-1.) An *unassigned area* is the area between noncontiguous areas of operations or beyond contiguous areas of operations. The higher headquarters is responsible for controlling unassigned areas within its area of operations (FM 3-0). The type of AO affects the corps's requirement to resource the mission. The unit controlling the ground has responsibility for terrain.

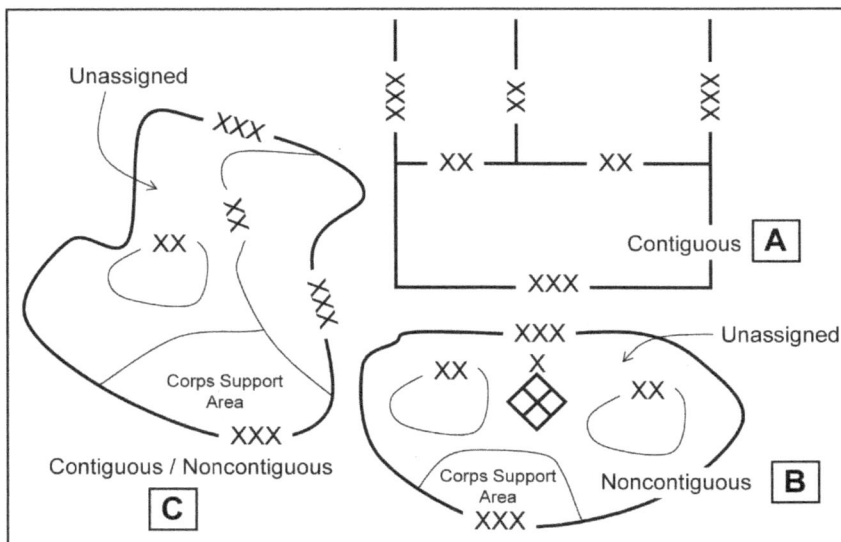

Figure 4-1. Possible configurations of corps areas of operations

Contiguous Area of Operations

4-11. In figure 4-1, the two divisions assign missions, set conditions, and influence the fight within their AOs. These divisions integrate and synchronize the warfighting functions. The corps headquarters supports the d ivisions' op erations an d prov ides th e in telligence, m ovement and m aneuver, fires, protection, sustainment, and other resources consistent with corps priorities. Sometimes no unassigned areas ex ist or the CG retains an unassigned area forward of the subordinates' AOs. In either case, the corps headquarters continues to observe the corps AO.

Noncontiguous Area of Operations

4-12. Figure 4-1 illustrates a corps AO divided into noncontiguous AOs. From the corps' perspective, this is more complex since the corps headquarters controls unassigned areas with in its AO as well as su pports division operations. The corps headquarters has responsibility of unassigned areas; it monitors those areas to counter any risks to the force.

Contiguous and Noncontiguous Mix

4-13. Also in figure 4-1 (page 4-2) is an a rea th at falls b etween t he two o ther options. Th e co rps headquarters adequately resources its two divisions, each with a different terrain management challenge.

4-14. A division with an AO contiguous to the corps support area benefits from sharing a boundary with its higher h eadquarters. A sh ared boundary can facilitate moving reso urces to sup port th e d ivision m ission. Likewise, th e d ivision can b enefit fro m th e in telligence, fires, protection, and sustain ment ac tivities exercised by the corps in the unassigned area. This division remains responsible for monitoring its portions of its area of influence that overlaps with the corps-controlled unassigned area.

4-15. The division in a noncontiguous AO s urrounded by t he cor ps-controlled unassigned area has t wo responsibilities. First, it co ordinates for support from the corps and for movement of resources between its AO and the co ntiguous areas of th e corp s AO. Secon d, it monitors act ivities in its area of i nterest and requests the corps to counter any threats originating outside the division AO but within its area of interest.

Unassigned Areas

4-16. The unassigned area is not em pty. At a minimum, civilians and possibly enemy ele ments live there. Regarding un assigned areas, corps actions aim to prevent the enemy from massing forces and capabilities that enda nger the corps. T he CG assess es the risk and reacts accordingly. In all or portions of the unassigned area, the CG may accept risk. When accepting risk, the CG has a branch plan to cover the areas with intelligence collection assets and sufficient forces to defeat a potential threat. Areas with a low risk of enemy occupation and action can be han dled in many ways. For exam ple, areas with no permanent force only receive periodic intelligence, surveillance, and reconnaissance (ISR) efforts. Other approaches include employing a full-time economy of force effort or assigning the mission to a corps major subordinate unit to deter a threat fro m intelligence collection, staging, or attacking from that area. At a min imum, the CG can assign "be prepared" tasks to subordinate units in the corps AO to anticipate operations and in telligence requirements.

4-17. The CG has several options available to g ain and maintain control of the in itially unassigned areas. The C G coo rdinates with jo int, ho st-nation, and m ultinational fo rces to tak e responsibility fo r all or portions of the area. The CG may elect to assign the entire area to parts of the corps, including having the division h eadquarters d ivide th e entire area. Control of i nitially u nassigned areas can be an essen tial or supporting task for the unit given the mission. For example, friendly forces always transit from contiguous to n oncontiguous areas or from o ne noncontiguous area to another. These fo rces can provide in telligence and coverage of t he area s. The forces r equired t o c ontrol a n u nassigned area differ de pending on t he situation, but often th ey will co nduct econ omy o f fo rce. When th e corp s con ducts decisive or sh aping operations, a small force or a small portion of a larger force then monitors and controls unassigned areas.

AVAILABLE CORPS FORCES

4-18. The nu mber of forces av ailable, th eir cap abilities, an d th eir ex pected du ration in th e co rps fo rce structure impact operational reach. In general, a mix of capable force s in the appropriat e numbers enables the corps to accom plish its o bjectives and achieve them before reac hing its culminating point. De pending on how the theater army tailors the forces to the corps, the CG has several forces available. These include a combination of one or more brigade combat teams (BCTs), support brigades (aviation, fires, maneuver enhancement, battlefield surveillance, and sustainment), functional brigades (including but not limited to engineer, chemical, military intelligence, and military police), and Army Reserves and its functional commands. Furthermore, the forces required to accomplish the mission come from more than Army sources. Manned and unmanned joint air assets and special operations forces (including special operations aviation) can support the corps mission.

Division Headquarters

4-19. The role of the division headquarters is to employ land forces as part of a joint, interagency, and multinational force during full spectrum operations. The division executes simultaneous offensive, defensive, and either stability or civil support operations (depending on whether or not it is operating in a foreign country or the United States) in an assigned AO to establish specific conditions. It combines tactical tasks and missions through its organization of decisive, shaping, and sustaining operations to accomplish its assigned mission. The division is the primary tactical warfighting headquarters for command and control of land force BCTs.

4-20. The trained and ready division headquarters supports the CG in exercising authority and direction of operations of subordinate BCTs and other brigades. Division headquarters subordinate to the corps facilitates flexibility and enables the CG to shape the operational environment.

Brigade Combat Team

4-21. Capabilities differ with each of the three BCTs (infantry, heavy, and Stryker); however, each can conduct sustained offensive, defensive, and stability operations in most environments. In support of corps operations, they train for small-unit operations, security missions, heavy-light integration, forcible entry, and early-entry operations. Each unit can conduct surveillance and reconnaissance operations. The heavy BCT and Stryker BCT can conduct strike operations. When assigned an AO, the BCT and armored cavalry regiment can provide ISR and protection required to monitor and counter or delay enemy action from unassigned areas. As powerful combat formations, they can defeat threats from their areas of interest alone or as a part of a larger force.

Maneuver Enhancement Brigade

4-22. Normally the maneuver enhancement brigades (MEBs) are assigned to, attached, or placed under the operational control (OPCON) of a division. However, they may be attached to or placed OPCON of a corps. The MEB is designed to control the following capabilities: engineers; military police; chemical, biological, radiological, and nuclear; and civil affairs. When provided to the corps, the brigade is usually assigned an AO, normally the corps support area, where it conducts terrain management, movement control, clearance of fires, security, personnel recovery, ISR, stability operations, area damage control, and infrastructure development. The MEB coordinates air and ground movement. The MEB can conduct route and convoy security operations for the corps or protect units as they move in movement corridors from one area to another. It coordinates with the theater distribution center to maintain visibility during movement operations within movement corridors and unassigned areas in the corps AO. The brigade's task to manage terrain in the corps support area may require it to establish a tactical combat force to counter threats to the support area from unassigned areas. (See FM 3-90.31 for additional information on the MEB.)

Combat Aviation Brigade

4-23. The combat aviation brigade provides reconnaissance, air attack, command and control, medical evacuation, and medium lift for corps operations. The brigade's digital connectivity lets it synchronize fires from supporting fires assets and ground maneuver forces with the brigade's own firepower to eliminate or suppress enemy threats. Depending on how it is tailored by the force provider or the theater army, the combat aviation brigade can support the corps's coverage of unassigned areas by aerial reconnaissance or a movement to contact. The brigade supports movements between noncontiguous AOs with heavy- and medium-lift and attack reconnaissance helicopters and with command and control assets. (See FM 3-04.111 for more information on the combat aviation brigade.)

Military Intelligence Brigade

4-24. The organization of the typical theater-level military intelligence brigade includes three tactical-level units: an operations battalion, a forward counterintelligence and human intelligence collection battalion, and a forward signals intelligence collection battalion. The military intelligence brigade contributes to the corps' ability to assess the threat in its AO by conducting intelligence collection and analysis, as well as developing and disseminating intelligence products.

Battlefield Surveillance Brigade

4-25. The organic ISR battalion of the battlefield surveillance brigade and support organizations in support of corps operations can be augmented as required. Augmentation can include additional military intelligence and ground maneuver units, aviation assets, unmanned aircraft systems, and engineer units. The brigade executes a corps-level ISR plan for the AOs and unassigned areas. The battlefield surveillance brigade enables the corps to exercise command and control over assets that collect against the corps's information requirements, including in the unassigned areas. The battlefield surveillance brigade headquarters can serve the corps as an intelligence fusion cell to pull in intelligence assessments and provide situational understanding.

Fires Brigade

4-26. Fires brigades are normally assigned to, attached to, or placed under the OPCON of a division but may be attached to or placed under OPCON to the corps. The fires brigade gives the corps a headquarters the ability to conduct close support fires, counterfires, and reinforcing fires across the corps AO. The brigade exercises command and control of Army and joint lethal and nonlethal fires. The fires brigade develops and recommends fire support coordination measures that enable it to support corps operations. These measures include fire support coordination lines, free-fire areas, kill boxes, no-fire areas, and restrictive fire areas. Depending on the situation, the corps CG may direct a commander of a supporting fires brigade to serve as a principle advisor or the force field artillery headquarters commander.

Engineer Brigade

4-27. The engineer brigade supports the corps, conducts engineer missions, and controls up to five mission-tailored engineer battalions, including capabilities from all three of the engineer functions. It also provides command and control for other non-engineer units focused on accomplishing missions such as support of a deliberate gap (river) crossing as needed. The engineer brigade headquarters design enables conducting operations to build local technical and engineering capacity.

Military Police Brigade

4-28. The military police brigade provides support to the corps as well as plans, integrates, and executes military police operations by up to five mission-tailored military police battalions and integrating capabilities from all five of the military police functions. It may also provide command and control for other non-military police units focused on accomplishing such missions as area support, internment and resettlement, or host-nation police development operations.

CONSTITUTE AND EMPLOY A RESERVE

4-29. The ability of the corps to command and control available forces closely relates to its ability to employ a reserve. Often operational reach depends on the combat power available later in an operation when the corps can reinforce a weak capability or exploit success. The CG relies on the staff and the main command post's (CP's) effectiveness when deciding on the corps reserve employment. Establishment, size, positioning, level of preparation directed, adjustment, movement, support, commitment, and replacement of the corps reserve all tie to the situation and the CG's situational understanding. These decisions stem from how the CG understands the operations process; visualizes the operations' evolution, potential branches, and decision points; and recognizes that events do not always go as planned. The main CP functional and integrating cells maintain the common operational picture (COP) so the CG can make decisions about the reserve and its e mployment. Th e m ain CP cells co ntinually assess th e situ ation, esp ecially d uring transitions from one operation to another.

4-30. The corps maintains a re serve based on the situation and its mission analysis. The C G determines how to commit and use the corps reserve, although the current operations integration cell can exec ute the order to commit the reserve when certain conditions are met based on the CG's guidance. The corps plans for the use of a reserve to deal with emergencies or expected contingencies, and, in the case of an offensive operation, to reinforce su ccess. Ofte n t he corps needs a reserve for offe nsive or defensive ope rations. However, stabilit y o perations m ay req uire a reserv e t o reinforce efforts to en sure civ il security an d civil control. Rar ely do ci vil suppo rt operations require a reserve, unless the incident or natural disaster co uld reoccur in another unexpected location requiring rapid response.

4-31. The CG can take risk and have no reserve. The CG can also identify one risk which to respond such as expected contingencies or emergency situations. The reserve needs enough combat power, mobility, and sustainment resources to accomplish its projected mission. The terrain or projected missions may require a division or more tha n one BCT. The CG assem bles the forces m ost likely to succeed in the c urrent operation. Li kely t hey consi st of a ground maneuver formation. Al though fire asset s are n ever kept i n reserve, they often receive on-order m issions t o support the reserve, such as with reinforci ng fires, suppression of enemy air defense, or blocking fires to shape the battlefield.

4-32. Sometimes, especially in stability and civil support ope rations, the CG may place a multifunctional or functional brigade in reserve. This unusual reserve works well since they have capabilities that facilitate all warfighting functions. However, the situation may call for potential reinforcement to ongoing operations with eng ineers, military p olice, ch emical defense, explosive ordn ance d isposal, quartermaster, g eneral aviation, or m edical forces. As with all re serves, all or part of a funct ional br igade receives an on-order mission to sup port an op eration. Ad ditionally, th e C G m ay k eep fun ctional brig ades i n reserve du ring offensive an d defensive operations t o o vercome unexpec ted ob stacles or p rotect t he cor ps f rom ai r and missile, chemical, unruly population, or health threats.

CONDUCT OF STABILITY OPERATIONS

4-33. The operational reach of the corps is enhanced by its ability to engage in all aspects of full spectrum operations. The corps' mission determines the rel ative we ight of ef forts am ong the o ffensive, de fensive, and stab ility or civ il sup port elem ents. The corps co ntinuously an d si multaneously p repares to con trol forces engaged across the spectrum of conflict. The corps condu cts operations to leverage the coercive and constructive cap abilities o f their force b y u sing t he app ropriate co mbination of defeat and stabilit y mechanisms that best acco mplish the m ission. The corp s headquarters contributes to estab lish co nditions that facilitate future success. (See FM 3-07 for further information on stability operations.)

FORCIBLE ENTRY

4-34. The operational reach of the corps is e xtended by *forcible entry*—seizing and holding of a m ilitary lodgment in the face of armed opposition (JP 3-18). Com manders typically use it in operations where the entry force either ca n hold on its own ag ainst the expected enemy force, or th ey anticipate a ground force can link up with the entry force to protect it and continue operations. Units can execute forcible entry via parachute, ai r assault, am phibious assa ult, or rapid m ovement ove r l and. A f orcible entry o peration i s inherently joi nt. Forci ble ent ry ope rations can use a si ngle method (s uch as ai r ass ault force s) o r be integrated (when combinations of early entry forces pa rticipate). If forces use more than one m ethod, their operations can be co ncurrent or i ntegrated. In c oncurrent ope rations, f orces e xecute di fferent t ypes o f operations simultaneously to accom plish differe nt object ives. In a n i ntegrated fo rcible entry, different capabilities seize different objectives, but the friendly forces are mutually supporting.

4-35. The corps headquarters trains to conduct Army and joint forcible entry operations. These operations may use any com bination of di visions o r B CTs an d s upporting units fr om multiple aeri al po rts of embarkation a nd sea ports of em barkation. The co rps headquarters, once d esignated a jo int task fo rce headquarters, may al so be r equired t o co nduct a joint fo rcible e ntry operation em ploying assa ult and support forces from the othe r Se rvices and the United St ates Special Operations C ommand. To res pond rapidly to m any contingencies, the corps headquarters prepares to echelon command and control facilities into the fight, starting with th e early-entry CP, the mobile command group, or the tactical CP. For large-scale operations, the main CP serves as the command and control headquarters.

4-36. The comm and and control headqua rters charge d with con ducting a fo rcible en try o peration establishes co mmand and s upporting rel ationships wi th the j oint, Serv ice, or fun ctional h eadquarters commanding and controlling the operation.

4-37. See JP 3-18 for additional information on the conduct of forcible entry operations.

CONSEQUENCE MANAGEMENT

4-38. The operational reach of the c orps expands with its ability to respond to results of combat actions or a disaster that requires a foc used res ponse by the co rps to re duce the s everity of t he im pact. Often, this limited intervention activ ity occurs with other combat actions in the corps AO. When this occurs, the CG decides whether the consequence management or the initial operation will be the economy of force effort.

4-39. To perform cons equence m anagement t asks, the c orps coordinates s upport t o m aintain o r restore essential serv ices an d m anage and mitig ate p roblems. These pro blems can resu lt fro m d isasters an d catastrophes, including natural, man-made, or terrorist incidents. Many incidents requiring a response fit in three categories: natural disasters; high-yield explosives; and chemical, biological, radiological, and nuclear hazards. Each incident requires a different response with varying military contributions.

4-40. Consequence management can be f oreign or d omestic. Forei gn c onsequence m anagement i s a stability operation. The host nation has responsibility of foreign consequence management; however, other nations m ay reque st U .S. assistance t hrough t he Department of St ate. Si nce the Department of Defense possesses many assets required to respond to a consequence management incident, the President may direct the De partment of Defense t o sup port t he Depa rtment of St ate o r ot her U. S. G overnment agenci es. Domestic consequence m anagement is a com ponent of civil support. Milita ry support for domestic consequence management normally falls to the Army and Ai r National Guard, which can be em ployed by civil au thorities wh ile serv ing in a Un ited States Co de, Title 3 2 statu s. The d eployed co rps h eadquarters with a consequence management mission encounters a level of risk. This risk occurs not only as a result of the disaster itself, but also becau se of a potential military threat. Sometim es consequence m anagement requires combat operations against an enemy to provide a secure environment permitting forces to perform consequence management t asks. Protecting jo int and mu ltinational m ilitary an d civ ilian fo rces and o ther responders requires constant effort. Often the need to respond requires the use or threat of force by friendly forces to create conditions for successful consequence management operations.

4-41. The corps headquarters can serve as a joint task force headquarters, serve as an ARFOR headquarters under a joint task force, or provide command and control for the consequence management operation in its own right. R egardless of i ts st atus, t he c orps hea dquarters plans, prepares, e xecutes, an d as sesses the response and recove ry operations in the AO. The c orps headquarters marshals resources of its subordinate units and m akes requests for available res ources from the theater army and national-level force providers. During f oreign co nsequence management si tuations, t he co rps s ubordinates i ts asset s t o assi st t he organizations of t he l ocal government. The m ain C P o rganizes i n t he sam e w ay fo r operations predominated by stab ility or civil sup port as it do es for combat operations. The functional and integrating cells ope rate according t o standing operat ing procedures, fulfilling the ir tasks in support of t he c orps command and control system.

4-42. FM 3-28.1 contains additional information on consequence management.

SYNCHRONIZING INFORMATION ACTIONS

4-43. Synchronizing information actions involves information engagement and supporting processes.

INFORMATION ENGAGEMENT

4-44. Information engagement (see paragraph 2-116 for the definition of information engagement) seeks to use th e power o f knowledge to in form an d in fluence internal a nd e xternal audiences. The c orps a nd its subordinates develop i ntegrated i nformation e ngagement task s with t heater arm y, j oint, and National strategic communications. Incorporating information engagement into the corps concept of operations is a staff responsibility of the assistant chief of staff, information engagement (G-7). With the assistance of the functional and integrating cells, the assistant chief of staff for information engagement coordinates the use of the information engagement capabilities: leader and Soldier engagement, public affairs, psychological operations, combat camera, and strategic communication and defense support to public diplomacy.

4-45. The corps focuses information engagement efforts within its AO. However, the efforts must be nested with those of higher and lower echelons to provide synergistic effects and not disrupt or confuse friendly forces rather than the target audience. Actions by corps Soldiers, both positive and negative, influence how the local populace perceives the military. Therefore, in all actions, leaders focus on managing expectations and informing the people about friendly intentions and actions. The assistant chief of staff for information engagement works with the assistant chief of staff, civil affairs operations (G-9). The latter normally works in the command and control cell to integrate civil affairs operations into the corps operations. The vehicle most often used for this is a civil affairs operations working group. It solves civil affairs operations problems and makes recommendations to the CG on how to incorporate civil affairs operations into corps operations.

SUPPORTING PROCESSES

4-46. The corps main CP uses information actions integral to the various processes to solve problems. Corps planning uses the military decisionmaking process to identify the problem, develop alternative solutions, subject them to thorough analysis, and make a recommendation to the commander. These results inform ISR activities, are developed into operation plans and orders, and provide the starting point to solve the next problem. The structure of the military decisionmaking process enables commanders to consider both lethal and nonlethal actions when developing corps plans and orders.

4-47. Information tasks are synchronized and integrated through activities such as the period synchronization meeting. With regards to targeting, the corps chief of fires leads the targeting working group and participates in the targeting meeting and joint targeting coordination board, when formed, in the corps main CP. Once the operation order is published and the operation controlled by the current operations integration cell, the daily operations synchronization meeting serves as a forum for final integration of lethal and nonlethal actions.

APPLYING THE ELEMENTS OF COMBAT POWER

4-48. The corps CG primarily provides resources. By applying the elements of combat power, the CG establishes priorities and provides resources to subordinates. Early in the planning process and throughout execution, the CG clearly articulates priorities of effort and support and identifies which units get resources in what order to accomplish the mission. The CG and staff identify the requirements to accomplish its mission and constantly communicate them to the force provider in the continental United States and the theater army. The corps headquarters establishes the objective, gives subordinate headquarters required resources for the current operation, monitors execution, and ensures that the force has the assets available to execute probable branches and sequels that will lead to or reinforce success. The CG applies the elements of combat power, allocates enablers, and lifts and shifts the main effort as required.

WEIGHTING DECISIVE OPERATIONS

4-49. Weighting decisive operations is the most direct means to resource for mission accomplishment. The CG can weight these operations by providing resources, setting priorities, shaping current operations, and planning future operations. In the corps operation order, the CG directs how corps forces are to cooperate. The concept of operations describes the commander's visualization. The CG synchronizes the operation so the main effort carries most of the combat power. The CG also establishes command and support relationships that provide immediate combat power and sustainment to the main effort. The ability of the CG to give the main effort the advantage in combat power depends on those assets tailored to the corps from the theater army or force generators in the continental United States.

4-50. Examples of corps actions to weight the main effort include the following:

- Task-organizing maneuver forces.
- Massing fires to support offensive operations. The corps can place one or more fires brigades in support of a BCT or division with priority of fires to the main effort.
- Nominating targets to the air component via the battlefield coordination detachment to employ joint fires and interdiction as part of corps shaping operations.
- Placing one or more internment and resettlement battalions to process the detainee population from an operation.
- Coordinating with the theater sustainment command to reinforce the main effort's sustainment organization (such as a BCT's brigade support battalion) with one or more sustainment brigades.
- Prioritizing sustainment support to the main effort to provide them mobility, munitions, maintenance, and other logistic activities.
- Placing additional engineers, military police, or civil affairs operations units under the OPCON of a BCT or division conducting a stability operation concurrently with an offensive operation.
- Placing aviation assets under the tactical control of a BCT or division to provide lift support for troop movement and resupply of critical items or ammunition.
- Attaching ISR units and assets—such as unmanned aircraft systems, ground cavalry, and military intelligence analysis capability—to the main effort.
- Placing psychological operations teams under OPCON of the main effort to conduct psychological operations before an operation.
- Attaching national assets, such as a forward contact team from the United States Army Medical Research Institute of Infectious Diseases, to provide vector analysis.
- Elevating priority of network resources, such as bandwidth and preemption level of information, dynamically enforceable by the network.

INTELLIGENCE, SURVEILLANCE, AND RECONNAISSANCE

4-51. The corps establishes priorities and allocates resources by apportioning enabling capabilities. Key among these is ISR. It enables the selected force to know the threat, the weather, and the terrain over which it must conduct operations. *Intelligence, surveillance, and reconnaissance* is an activity that synchronizes and integrates the planning and operation of sensors, assets, and processing, exploitation, and dissemination systems in direct support of current and future operations. This is an integrated intelligence and operations function. For Army forces, this activity is a combined arms operation that focuses on priority intelligence requirements while answering the commander's critical information requirements (FM 3-0). The main CP controls ISR. The corps sets the conditions for successful mission accomplishment by providing the lowest levels of the corps with the assets necessary to gain and report required information.

4-52. In the corps AO, the main CP has responsibility for intelligence. Subordinate divisions, functional brigades, and BCTs perform surveillance and reconnaissance tasks to collect information on the enemy, terrain, weather, and civil conditions. The corps requires intelligence to maintain situational understanding, support targeting, and facilitate information engagement. To accomplish these tasks, the CG directs the staff how to plan for, provide, and employ collection assets and subordinate forces. Surveillance is a continuing task; it is not oriented to a specific target. It is designed to provide warning of enemy initiatives and threats and to detect changes in enemy activities. Reconnaissance complements surveillance by obtaining specific information about activities and resources of an enemy, potential enemy, or geographic characteristics of a particular area.

4-53. Corps ISR operations vary in collection techniques in full spectrum operations. The ability to collect information, provide competent analysis, and exploit it as rapidly as possible, act on it at a measured pace, or not to act on it at all, is the key to mission accomplishment. A multidivisional corps force fighting a similar enemy requires applying ISR assets differently from a fragmented counterinsurgency operation. At the high end of the spectrum of conflict, a centralized approach to ISR synchronization and integration often differs from a decentralized approach at the low end. The CG determines the approach based on the situation.

4-54. The main CP directs ISR efforts. As directed, corps subordinate forces conduct ISR activities. Intelligence activities receive directions from the corps operation order and operations synchronization process. Corps-controlled collection assets are allocated to subordinate forces. Results of the corps-wide collection and preliminary analysis efforts are passed to the main CP. The main CP centers ISR synchronization in the intelligence cell and ISR integration in the movement and maneuver cell.

4-55. ISR synchronization entails analyzing information requirements, identifying intelligence gaps, and developing commander's critical information requirements. It results in an intelligence plan and requests for information to corps subordinate organizations to fill knowledge gaps. ISR assets collect information from all sources, provide preliminary analysis, and develop the situation. ISR forces normally conduct economy of force so that most corps forces avoid contact with the enemy without adequate intelligence.

4-56. The assistant chief of staff for intelligence, with input from all functional and integrating cells, oversees corps ISR synchronization. All staff elements in the intelligence cell contribute to the effort by providing subject matter experts. The intelligence cell, with the help of the communications integration element and others, designs the intelligence architecture so that it can pass information rapidly. The intelligence cell's ISR operations element develops priority intelligence requirements with input from the entire staff and requests for information. The intelligence cell refines and pairs the requirements with collection means. The intelligence fusion element receives, processes, analyzes, and disseminates intelligence. Other cells participate. The fires cell contributes to ISR synchronization during the targeting process as they detect potential targets for lethal or nonlethal activities. These activities include identifying potential electronic warfare targets. The staff weather officer describes weather effects. Such identification enables the staff to further determine how weather may affect collection activities and how weather impacts their areas of expertise.

4-57. ISR integration requires assigning and controlling the corps's ISR assets with regards to time, space, and purpose. In coordination with the intelligence cell, the assistant chief of staff, operations (G-3) integrates corps ISR activities. The intelligence cell's ISR operations element integrates intelligence products and collection planning into current operations. It is the interface between the intelligence cell and the movement and maneuver cell. The ISR operations section coordinates the collection effort across all functional and integrating cells matching tasks with required assets. The ISR operations section's target development element ensures that targets are developed, prioritized, and sequenced into current and future operations.

4-58. The corps can push ISR assets to the lowest tactical level. The CG can weight the ISR effort with assets from theater army and corps forces based on commander's critical information requirements. Potential units for ISR collection are battlefield surveillance brigades, combat aviation brigades, maneuver enhancement brigades, and reconnaissance units of subordinate divisions and BCTs. These forces combine with national- and strategic-level collection platforms to fill information and intelligence requirements. Decentralized ISR collection assets, including providing BCTs with unmanned aircraft systems, give the lowest tactical-level imagery and signals intelligence support. Control of ISR assets at the lowest level possible is the key to adequate and timely intelligence at corps level.

4-59. See FM 2-0 for additional information on ISR operations.

UNMANNED AIRCRAFT SYSTEM SUPPORT

4-60. The unmanned aircraft system (UAS) is a force multiplier for the corps. The joint UAS normally focuses on theater-wide intelligence gathering and surveillance activities. Commanders can focus a UAS on reconnaissance. As an Army intermediate tactical headquarters, the corps cannot always rely on full-time joint UAS availability to support its operations. Even if allocated to support corps operations, higher priority missions may divert a joint UAS mission before friendly forces begin or prior to mission accomplishment.

4-61. The UAS supports corps in several ways. Army UAS provide direct support to ongoing operations. The UAS provides surveillance, reconnaissance, attack, communications relay, and convoy overwatch. The UAS acquires the enemy force, and either keeps it under observation or hands it over to aviation or ground assets for continued observation or destruction. After the UAS receives targets from ground maneuver or manned aircraft, it continues to observe or engage with organic fires. The UAS also facilitates engagement by other assets such as field artillery, attack helicopters, close air support, or ground maneuver. Able to loiter longer than helicopters, the UAS assists in continued intelligence collection and battle damage assessment. By prioritizing UAS assets, the corps main CP can extend its reach beyond the limited or ground-based systems.

4-62. UASs can locate and identify targets by day or night and during reduced visibility to provide real-time surveillance by data-linked, electro-optical, or infrared sensors: They also can provide laser designation of targets for attack.

Unmanned Aircraft System Platforms

4-63. The corps has access to the following UAS platforms:

- Raven.
- Shadow.
- Hunter.
- Extended range multipurpose.

4-64. The Raven small unmanned aircraft system (RQ-11B) provides a small unit with enhanced situational awareness and increased force protection by providing expanded reconnaissance and surveillance coverage of marginal maneuver areas. Raven is a hand-launched and rucksack portable UAS. It consists of three air vehicles, a ground control station and remote video terminal, electro-optical and infrared payloads, a ground antenna, a field repair kit, and one initial spares package. It can fly for 90 minutes with a range of 10 kilometers. As a small UAS operating at the same altitudes as manned aircraft, the Raven creates challenges in airspace coordination and deconfliction.

4-65. The Shadow-200 tactical unmanned aircraft system (RQ-7B) is the Army's current force UAS for the BCT. The Shadow system provides Army brigade commanders with tactical-level reconnaissance, surveillance, target acquisition, laser designation, battle damage assessment, and communications relay. It is catapulted from a rail launcher, lands via an automated take-off and landing system and is supported by a platoon of 22 Soldiers. The Shadow system provides over 6 hours endurance and can be operated at 120 kilometers. It has an early-entry configuration that can be transported via three C-130s (Hercules).

4-66. The Hunter unmanned aircraft system (MQ-5B) is a multimission system which provides ISR, target acquisition, and battle damage assessment capability to division and corps commanders. The modular Hunter system uses the tactical command data link. Such systems enable the Hunter to be tailored to the specific location and mission requirements, including electro optical infrared laser designator, communications relay, Greendart, and weaponization. Hunter provides 18 hours endurance and can be operated at 200 kilometers.

4-67. The extended range multipurpose system (MQ-1C, sometimes known as Warrior) provides the division commander with a dedicated, assured, multimission UAS for the tactical fight assigned to the combat aviation brigade in each division and supports the division commander's priorities. The extended range multipurpose system provides reconnaissance, surveillance, target acquisition, command and control, communications relay, signals intelligence, electronic warfare, attack, detection of weapons of mass destruction, and battle damage assessment capabilities. A company of 128 Soldiers within a combat aviation brigade operates and maintains the system. The extended range multipurpose system can operate beyond the line-of-sight at distances greater than 300 kilometers.

Unmanned Aircraft System Contributing Cells

4-68. Functional and integrating cells contribute to synchronize and integrate UAS operations. The movement and maneuver cell contains seven elements, two of which contribute to UAS. The aviation element coordinates and synchronizes UAS activities in corps planning and monitors UAS operations to deconflict operations with other airspace users. Deconfliction requires constant attention. The airspace over the corps AO can become crowded with unmanned aircraft, manned rotary- and fixed-wing aircraft, and indirect fires. The airspace command and control element develops and coordinates the airspace control architecture during planning. It develops the required airspace coordinating measures and fire support control measures and monitors operations for compliance.

4-69. The intelligence cell provides tasks for UASs, focusing on intelligence collection. The intelligence collection management element monitors UAS collection activities and integrates and synchronizes UAS use to satisfy commander's critical information requirements. The intelligence signals element develops tasks to optimize collection of electronic intelligence. The imagery intelligence element receives, processes, and disseminates UAS imagery to meet information requirements.

4-70. The fires cell develops targets suitable for UAS attack resources and integrates them into the fire support plan. The fires cell provides input to the ISR plan to synchronize it with regards to designated targets. The field artillery intelligence officer coordinates with the intelligence cell for target selection, prioritization, and assessment. The electronic warfare element coordinates with the intelligence cell to synchronize electronic warfare and counter-electronic warfare activities for current operations and plans. The fires cell coordinates clearance of fires with other cells and elements in the main CP.

4-71. The protection cell coordinates with other functional and integrating cells. Together they conduct the Army support to joint personnel recovery and corps protection operations facilitated by UASs.

4-72. The sustainment cell coordinates with the intelligence cell for appropriate sustainment support such as repairs, parts, and maintenance of the UAS.

4-73. The command and control cell is the lead organization in the main CP. It ensures the necessary communication networks are in place and maintained during UAS operations. The cell coordinates with the other functional and integrating cells to synchronize network management, communications security, and information assurance into UAS operations. Depending on the situation, the network management element supports UAS communications relay activities. The command and control cell coordinates UAS actions that affect psychological operations, civil affairs operations, and information engagement activities in the corps AO.

4-74. Depending on the situation, the ISR operations section can be given tasking authority and tactical control over Army UAS in the corps AO. In coordination with the functional cells, the ISR operations section integrates UAS activities with the corps concept of operations and directs lethal and nonlethal actions. When the UAS ground control station exercises tactical control, the ISR operations section and other current operations integration cell support sections monitor the situation and provide necessary support.

4-75. For additional information on UASs, see FM 3-04.15.

TARGET ACQUISITION

4-76. Resourcing for target acquisition enhances corps operational capabilities. Without knowledge of the enemy and its intentions, the corps fire support and other systems cannot contribute to mission accomplishment, or worse will spend its resources on unproductive or counterproductive targets. *Target acquisition* is the detection, identification, and location of a target in sufficient detail to permit the effective employment of weapons (JP 3-60). At the corps level and below, ground reports commonly provide target acquisition. Soldiers on the battlefield observe the situation as they perform a task. They are augmented by human intelligence. Scouts, reconnaissance patrols, observation posts, long-range surveillance units, detachments left in contact, artillery observers, combat observation and lasing teams, and fire support teams at battalion and BCT levels provide this intelligence. Ground observers receive assistance from remote electronic and acoustic sensor systems. Electronic systems augment human intelligence by using the electromagnetic spectrum to detect information in the AO. Manned and unmanned aircraft support target acquisition.

4-77. Target acquisition information is received, processed, analyzed, and disseminated at the main CP in the intelligence and fires cells. In the intelligence cell, several sections coordinate and help synchronize target acquisition, including the ISR operations and G-2X (counterintelligence and human intelligence operations) sections. The current operations integration cell integrates target acquisition activities into day-to-day operations.

CIVIL SUPPORT OPERATIONS

4-78. In the continental United States, the Army National Guard, Active Army, and sometimes Army Reserve, civilians, and contractors work together to conduct civil support operations. Proper resourcing of the corps facilitates this part of corps operations. The corps can be called upon to interact with civil authorities. *Civil support* is Department of Defense support to U.S. civil authorities for domestic emergencies and for designated law enforcement and other activities (JP 3-28). Army forces provide this support when requested and private, local, state, and other federal government resources are insufficient to protect the life, limb, or property of citizens. Most Army civil support operations are conducted by the state-controlled Army National Guard. Generally, the Active Army is called upon when the Army National Guard requires augmentation for a disaster or other incident response.

4-79. The Army's roles and responsibilities for civil support operations fall under four primary tasks:
- Provide support for domestic disasters.
- Provide support for domestic chemical, biological, radiological, nuclear, or high-yield explosives incidents.
- Provide support for domestic civilian law enforcement agencies.
- Provide other designated support.

4-80. For more information about Army civil support operations, see FM 3-28.

THE CORPS ROLE IN THEATER AIR AND MISSILE DEFENSE

4-81. Air defense and its twin air superiority are always a consideration of the joint force commander. The enemy may be able to strike from the air with rockets, ballistic, and cruise missiles, fixed- and rotary-wing aircraft. However, the enemy's capabilities, or lack of capability, and UASs may enable the theater commander to accept risk and establish theater air and missile defense (TAMD) as an economy of force effort.

4-82. TAMD falls under the control of the area air defense commander. Normally, the area air defense commander is the joint force air component commander or a senior officer reporting to that commander. The area air defense commander integrates all aspects of air and missile defense (AMD) in the joint operations area. Specifically, the area air defense commander contributes to force protection through the suite of command and control systems, sensors, and shooters. TAMD systems provide information on the threat from the airspace to facilitate situational awareness and the COP.

4-83. The centralized approach of the TAMD fight commanded and controlled by the area air defense commander includes both Air Force and Army command and control, sensor, and strike assets in its decentralized execution. The theater army's senior air and missile defense headquarters is the Army air and missile defense command, a theater-level organization to which subordinate air defense units are assigned. Normally its commander is the deputy commander for the area air defense commander. Depending on the situation, these units can include air defense brigades, air defense battalions, and air defense batteries. The battalions are equipped with Patriot antiaircraft and antimissile units and short-range air defense systems such as the Avenger. The battalions can be all Patriot, all Avenger, or a mix depending on the threat. Separate air defense artillery batteries may include terminal high-altitude air defense and joint elevated networked sensor units. Because the threat from aircraft is much reduced, most air defense units concentrate on ballistic missile defense with some units identified as counterrocket, artillery, and mortar formations.

4-84. The corps main CP contributes to the TAMD effort by coordinating the TAMD assets provided by the theater army. The corps main CP's protection cell and its AMD element integrate TAMD assets into the corps plans and operation orders. The air defense brigade commanders advise the CG on counterair. These commanders work closely with the AMD element on synchronizing, integrating, and employing AMD capabilities. The corps AMD element plans, provides early warning, recommends asset allocation, develops the defended and critical asset list, and works with other main CP cells to coordinate airspace. The corps AMD tasks focus on these objectives:

- Ensuring freedom of maneuver by eliminating the air and missile threat.
- Achieving information superiority by collecting, processing, and disseminating airspace information.
- Protecting corps assets from attack.
- Protecting geopolitical assets—those friendly or host-nation locations deemed for priority protection.

Chapter 5

Corps Headquarters Transition to a Joint Task Force Headquarters

Army transformation aims to improve the Army's ability to provide flexible and responsive capabilities to joint force commanders. The corps headquarters primarily serves as an intermediate Army land force headquarters. This chapter describes the process as the corps headquarters transitions into a joint task force headquarters or as a joint force land component command headquarters. It discusses the joint force, the transition to training cycles, the joint task force headquarters organization, augmentation, and joint land operations.

THE JOINT FORCE

5-1. The Army corps headquarters uses joint doctrine and procedures when serving as joint force headquarters. As such, it operates in an environment with its own lexicon as described in Joint Publication (JP) 3-33. Army commanders must know and understand joint terms in addition to Army terms. Table 5-1 depicts several terms that the corps will encounter when transitioning to a JTF headquarters, and some are used to describe this process below.

Table 5-1. Common joint terms

Term	Definition
Service component headquarters	A combat force that is organized, manned, equipped, and trained to perform Service and functional roles.
designated Service component headquarters	A Service headquarters selected by the geographic combatant commander to be trained and serve as a joint capable headquarters.
joint task force (JTF)-capable headquarters	A designated Service headquarters that is certified and reports its readiness to perform as a joint headquarters.
joint task (JTF) force headquarters	A headquarters designated by the Secretary of Defense, geographic combatant commander, subunified commander, or an existing joint task force commander to conduct military operations or support to a specific situation.
standing joint force headquarters (core elements)	A full-time joint command and control element that is part of the geographical combatant commander's staff and focuses on contingency and crisis action planning.

5-2. The Army provides forces and capabilities to the joint force commander through the Army force generation process. When designating the corps as a joint task force (JTF) headquarters organization, the combatant commander expects a trained and ready force. Often corps forces demonstrate more proficiency in Army operations than in joint operations. Army forces train using a three-phased readiness cycle that mirrors the Army force generation cycle. Training and readiness in the Army force generation cycles around the reset, train/ready, and available phases. In concept, the reset phase contains units returned from a deployment or window of opportunity for deployment. Units perform those recovery, reconstitution, and minimal training activities required to become ready for future operations. Reset focuses on the Soldier, the Soldier's family, and equipment. Units in the train/ready phase conduct individual and collective training and perform other tasks necessary to prepare for deployment. While training, train/ready phase units still support civil authorities for national emergencies. The command still might call on the units to conduct a second major contingency (though they would not be as ready as if they were in the available pool). Finally, units in the available phase, including corps headquarters, deploy as required or are available for contingency operations. Units in the available force phase can deploy worldwide. This is the highest state of readiness. While this is an Army concept, the joint force commander directs the incorporation of joint training to ready the corps headquarters for immediate use.

THE TRANSITION TRAINING LIFE CYCLE

5-3.   The combatant commander may maintain a standing joint force headquarters (core element) fully integrated into the planning and operations, which can quickly form the nucleus of a JTF headquarters. The end state is not to have a JTF headquarters for every conceivable contingency, but to have one or more Service headquarters certified through pre-crisis training to respond when called. Pre-crisis training aims to enhance basic capabilities and skill sets resident in the Service component. These capabilities and skill sets enable the headquarters—in this case an Army corps—to function with augmentation as a JTF headquarters. After training and certification, the JTF-capable headquarters has a foundation of staffing, training, and equipment to serve as a joint headquarters.

5-4.   For an Army corps headquarters, the transition training life cycle contains five phases: preparation, certification, activation, employment, and reset. (See figure 5-1.) The phases represent the combined actions required to organize, staff, equip, train, and certify the corps headquarters as joint capable.

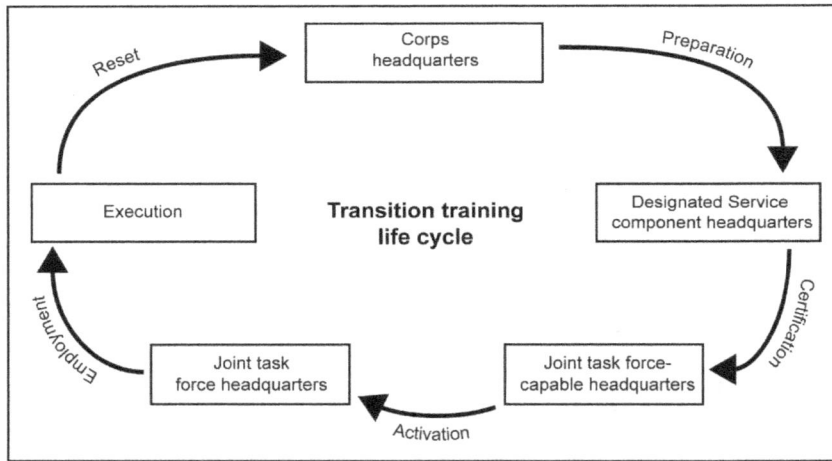

Figure 5-1. Transition training life cycle

PREPARATION PHASE

5-5.   The preparation phase for a corps headquarters begins simultaneously with the Army force generation train/ready phase. Training a corps headquarters to become joint capable may prove intensive. Regardless, a corps headquarters retains its capacity to perform the required Army-specific tasks. Its designation as a Service headquarters does not relieve the corps from maintaining an appropriate C-level for equipment and training to accomplish its wartime Army missions as directed in Army Regulation (AR) 220-1.

5-6.   The corps headquarters also trains on joint tasks such as those contained in the Universal Joint Task List (see the References for the Universal Joint Task List Portal Web address; for additional information, refer to Chairman of the Joint Chiefs of Staff Manual (CJCSM) 3500.04E). The combatant commander directs or identifies these tasks as the corps headquarters conducts its mission analysis and readiness training. Regardless of any joint designation, the corps headquarters remains available to deploy in the continental United States for civil support operations.

5-7.   The Army corps headquarters develops a joint mission-essential task list for approval by the joint commander. Concurrently, the designated Service headquarters develops two other documents: the joint manning document (JMD), and a joint mission-essential equipment list. The JMD and its companion joint mission-essential equipment list implicitly guide the corps commander and staff as they develop, approve, execute, and assess the training plan to accomplish the mission.

5-8.   As the joint force provider, United States Joint Forces Command has developed three JMD templates to address likely missions: civil support operations, stability operations, and major combat operations. These templates serve as the starting point for mission-specific JMDs. Figure 5-2 shows the templates address manning needs of likely missions. The core element JMD serves as the base with additional augmentees to support the specific mission. Once designated as the core of a JTF, the corps trains on the core element tasks and, based on guidance and the geographic combatant commander's intent, focus the training effort on one of the three joint missions. These documents guide the corps headquarters and enable it to establish a training plan, a battle roster (including required augmentees), and the equipment necessary to operate in a joint and combined environment.

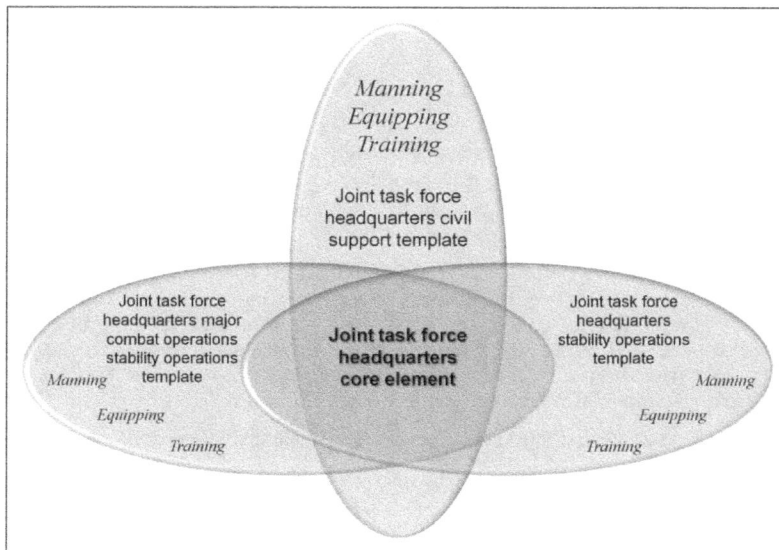

Figure 5-2. Joint task force augmentation templates

5-9.   Unit equipment varies. A joint mission-essential equipment list begins with the designated unit developing a thorough mission analysis. This equipment list contains information on the required equipment available in-theater and identified as shortfall. With the theater army or Army Service component command and the joint community, the designated Army JTF headquarters works to make up shortfalls through requisitions and redistribution.

5-10.   The preparation phase ends when a corps headquarters is identified as a designated Service headquarters.

CERTIFICATION PHASE

5-11.   The certification phase begins when the corps headquarters is identified as a designated Service headquarters.

5-12. Certification is based on Department of Defense policies that direct the JTF-capable headquarters to develop joint mission-essential tasks and provide periodic reports on readiness. Reports usually occur every 30 days or when the readiness posture of the JTF-capable headquarters changes. To facilitate certification, the certifying commander identifies possible missions, so the corps headquarters can concentrate its resources on tasks based on the certifying commander's intent. The certification process consumes resources, requiring a commitment from both the combatant-level commander and the designated unit. The certifying headquarters determines the length of the training and the time the designated JTF-capable headquarters will remain certified. The corps headquarters, when designated as a JTF-capable headquarters makes the initial certification step by conducting a thorough mission analysis to determine what the directing headquarters has told the corps headquarters to do. This analysis enables the corps to focus on those missions most aligned with the geographic combatant commander or joint force commander and the most likely circumstances for employment. It can isolate those areas that require the most training time and resourcing help from the geographic combatant commander, the potential theater army, the ARFOR, or others, especially the joint force enablers discussed in paragraphs 5-60 through 5-68.

5-13. Certification requires close coordination between the directing joint force and the corps headquarters. As training proceeds, draft documents—joint mission-essential tasks, JMD, joint mission-essential equipment list—are submitted to the certifying headquarters for review and approval. Close coordination enables the corps to identify shortfalls and develop strategies to address them. Although criteria for certification depends on the mission, the joint force commander (JFC) identifies several key certification standards:

- Competency in the JTF headquarters core joint mission-essential tasks.
- Competency in other mission-related tasks.
- A valid and resourced JMD, to include adequate liaison to sister Services, other government agencies, and, if required, multinational entities.
- A valid and resourced joint mission-essential equipment list.
- Acceptable mission readiness posture as measured by the unit status report or the Department of Defense Readiness Reporting System.

5-14. Certification training includes field training, orientation visits, joint command post (CP) exercises, and other assessment events. Because all required augmentation of personnel and equipment may not be available for every exercise, training exercises may occur in discrete increments. Simulation-driven exercises must stress the ability of the corps to accept augmentation, operate and maintain joint-capable equipment, and demonstrate its understanding of the missions. The corps headquarters can expect a graduation exercise as a final check on its mastery of the required skills.

5-15. The end state of the certification process is the designation of a corps headquarters as a JTF-capable headquarters.

ACTIVATION PHASE

5-16. The activation phase starts when the corps headquarters is formally designated as a JTF-capable headquarters by a combatant commander.

5-17. Once activated for a mission by the combatant commander, the joint and the corps commanders are jointly responsible for maintenance of a positive readiness posture as the corps and its apportioned forces prepare for action. This is accomplished by monthly readiness reporting through both Army and joint channels, continued execution of the joint training plan, periodic training to sustain both individual and collective skills, refinement of the JMD and joint mission-essential equipment list, and joint training exercises to keep unit skills sharp. The key to success of this continuous process is constant information sharing to maintain situational understanding.

5-18. The corps headquarters uses joint doctrine when preparing for deployment as a JTF in support of a combatant commander. As forces assemble and conduct pre-deployment exercises, the combatant commander monitors and mentors the process. The corps commander and staff receive, integrate, and train with the augmentees in live, virtual, or constructive training exercises. In most cases, this training occurs for the first time with personnel and equipment augmentation. Assuming the mission profile is known at the time of activation, the joint mission-essential task list, refined for the mission at hand, drives training. The joint mission-essential task list is both directed by the combatant commander and derived from doctrinal sources, such as CJCSM 3500.04E and JP 3-33.

5-19. The activation phase ends when the corps headquarters is trained and ready to serve as a JTF headquarters with adequate personnel and equipment augmentation.

EMPLOYMENT PHASE

5-20. The employment phase begins when the corps headquarters and its advanced echelon deploys to the joint operations area. As a subordinate of the combatant command in a joint operations area, the corps headquarters serving as a JTF headquarters exercises command and control over operations using joint doctrine. The employment phase ends when the mission is accomplished and residual operations are passed to a follow-on joint, multinational, or host-nation forces.

RESET PHASE

5-21. The reset phase starts when the corps headquarters redeploys to home station or another location to recover from the operation. The reset phase of the transition training life cycle is the same as the reset pool activities in as described in Field Manual (FM) 7-0 with its focus on the rehabilitation of the Soldier, the family, and equipment. The reset phase ends when the corps headquarters again joins the train/ready pool.

JOINT TASK FORCE HEADQUARTERS ORGANIZATION

5-22. The JTF commander exercises command and control. The properly designated JTF commander exercises authority and direction over assigned and attached forces to accomplish the mission. The commander assesses the situation, makes decisions, and directs action. The joint staff performs functions that help the commander exercise command and control responsibilities, including providing information, performing analysis, preparing plans and orders, and providing assessments on the progress of operations.

5-23. If designated to serve as the core of a JTF headquarters, the corps headquarters transforms into a joint staff. The corps headquarters organizes with its functional cells based on the warfighting functions and its integrating cells transitioned into a joint staff with its similar but more complex mixture of functional and integrating elements. During this transformation process, the assistant chief of staff, plans (G-5) and plans directorate of a joint staff (J-5) force integration section coordinates with other staff sections to maintain visibility of the process. It prepares the JTF headquarters as it grows in size, receives new capabilities, and becomes a truly joint organization. To facilitate this process, the corps staff, with augmentation, uses joint doctrine for staff functions and activities, referring to joint publications, Chairman, Joint Chiefs of Staff directives, and materials found in the Joint Electronic Library for direction and guidance.

5-24. Organization of the JTF headquarters includes—
- Staff organization.
- Joint staff augmentation.
- Joint Force headquarters formation.
- Reception, integration, and training.

STAFF ORGANIZATION

5-25. The JTF command group is similar to that of an Army organization with a commander, one or more deputies, a senior enlisted advisor, and several aides and personal assistants. The command group is assisted by personal and special staff groups to handle special matters over which the JTF commander wants to exercise personal control. Typical members of this group include a public affairs officer, staff judge advocate, chaplain, inspector general, command surgeon, provost marshal, safety advisor, and political advisor. This group may expand to fit the circumstances—for example, personal interpreters or translators, a cultural advisor, and special liaison officers. The staff is led by the chief of staff who supervises staff actions and serves as the principal integrator. Established and managed staff processes and procedures support the command's decisionmaking process. Typically reporting through a deputy commander to the JTF commander, the chief of staff oversees organizational integration, efficiency, and effectiveness. (See JP 3-33 for additional information on the organization and function of the joint staff.)

JOINT STAFF AUGMENTATION

5-26. Individual and unit augmentation to the JTF is crucial to operational readiness. Augmentation for each JTF staff section is available from the various agencies and organizations that support the joint community, including the theater army, theater-level commands, other geographic combatant commander Service commands, and national agencies. Each staff section has two key tasks for mission analysis. The first is to forecast the required fill of its battle roster. The second task is to identify vacancies that must be filled from external sources. To be most effective, augmentation requires a quick fill of slots in a joint manning document or other force generation mechanism. Team building and a thorough understanding of the specific mission and associated tasks are facilitated when the team comes together early and trains together in all preparation activities.

JOINT FORCE HEADQUARTERS FORMATION

5-27. As the primary U.S. land component, the designated Army joint force headquarters is often required to respond to crisis situations on short notice. The deployments can be as small as a no-notice permissive noncombatant evacuation operation during unstable peace or as broad as a general war involving a large multinational coalition. The corps headquarters provides a base structure on which to build a joint force headquarters for a small contingency with minimal additional Army augmentation, usually when most forces involved are land units. To serve as a joint force headquarters, the corps headquarters will always require joint augmentation.

5-28. As a joint force headquarters, the corps headquarters uses its own information systems, such as the Army Battle Command System (ABCS). It also confronts a numerous joint and multinational command and control information systems and their associated databases. Integrating these systems with corps information systems presents training, technical, and operations security challenges for the corps headquarters as it continues to effectively exercise command and control. A significant part of the pre-deployment preparation in the CPs involves training in using these systems.

5-29. Often, the technical integration of differing command and control information systems presents a greater challenge since forces do not share databases and operating systems across technical boundaries. With hundreds of potential applications and thousands of potential users, the staff establishes and monitors the corps and joint network infrastructure architectures for reliability. Potentially many commercial business and command and control systems compete for the same bandwidth. An overreliance on an untrustworthy system can lead to command and control problems if the system fails and hampers a commander's ability to lead the force. Likewise, operations security is an important consideration, especially for systems using an air gap to transfer voice and data from one command and control node to another. Commanders and staffs understand the levels of security to understand the security classifications of all the systems resident in the corps CPs.

RECEPTION, INTEGRATION, AND TRAINING

5-30. The appropriate number of augmentees in the correct slots may not contribute fully to mission accomplishment without proper reception, integration, and training. To facilitate this effort, the operational plans and interoperability directorate of a joint staff (J-7) has developed a series of joint education and training publications to integrate and train those assigned to a joint force headquarters. These publications describe the joint training system for the entire joint community. As described in Chairman of the Joint Chiefs of Staff Guide (CJCS Guide) 3501, the joint training system stems from these guiding principles:

- **Use joint doctrine.** It provides the basis for education and training as well as describing the employment means of U.S. forces in support of national ends.
- **Use commanders as primary trainers.** Commanders are responsible for preparing their commands to accomplish assigned missions.

- **Focus on the mission.** Training focuses on mission-essential tasks of the command in support of assigned missions.
- **Train the way you intend to fight.** As much as possible, make conditions and standards resemble those expected during a deployment or other mission thorough live, virtual, or constructive training.
- **Centralize planning and decentralize execution.** With overarching training objectives always in mind, the training and exercise program of the command trains every echelon from the commander to the Soldier.
- **Link training assessment to readiness assessment.** The command must be fit to fight and all aspects of the training and exercise program must be measured to determine organizational readiness.

5-31. Establishing a joint reception center facilitates the reception, initial processing, accountability, onward movement, and integration of replacements, augmentees, contractors, and others. Normally the responsibility of the manpower and personnel directorate of a joint staff (J-1), the joint reception center coordinates with the logistics directorate of a joint staff (J-4) for billeting, transportation, food service, medical support, and other requirements for newly arrived personnel. In one location or dispersed, the joint reception center accounts for multiple entry and exit points into the joint operations area. It provides such things as orientations, briefings, religious support, initial billeting, joint training, onward movement of units or personnel, and accountability of all personnel joining the JTF. Briefings can cover rules of engagement, rules for the use of force, cultural concerns, safety, operations security, and familiarization with JTF headquarters, dining areas, and other facilities. See JP 3-33 and JP 1-0 for additional information on the joint reception center. For strength accountability, the joint reception center is equally important as an outprocessing center.

5-32. The staff principal, with the help of the JTF joint reception center, has responsibility to orient and train each staff element. For ease of integration, the joint reception center should be staffed with JTF personnel from all Services to handle Service-specific requirements.

5-33. The corps headquarters staff and battle-rostered augmentees have integration and training opportunities to master the tasks identified in JP 3-33. The joint force command's knowledge development and distribution capability uses Internet-based distance learning. This learning prepares staff and augmentees for joint duty before and during deployment, exercise participation, and collective training. Its courses serve to orient those with limited knowledge of joint operations and reinforce previous training. This self-study combines with the periodic augmentee training provided by the geographic combatant commanders and others to ground the students in joint doctrine and practice. Individual training efforts expose students to joint doctrine, tactics, techniques, and procedures as well as subject matter experts. Such efforts enable students to develop a knowledge base so they can more quickly interact with joint staff colleagues.

5-34. The foundation built during individual training is further enhanced with a series of collective training events. Some events are previously scheduled or part of a predeployment readiness program. The Joint Chiefs of Staff sponsors CP exercises, mission readiness exercises, mission rehearsals, and staff assistance visits from the joint staff, JFC, and others. Such training reinforces learning, builds cohesion, and generates lessons learned. The corps headquarters designated as a potential JTF headquarters can expect to execute these collective training exercise in a combination of training environments:

- Live (real people in real locations using real equipment).
- Virtual simulation (real people in a simulation-driven situation).
- Constructive simulation (wholly simulated).

The combination of each of these environments creates a more realistic training environment for the corps headquarters.

5-35. A non-unit augmentee or deploying joint enabling capability member can join a JTF deploying to an area of operations outside the continental United States for an exercise or an operational deployment. This member completes the pre-deployment process and training for overseas service at a home station, continental replacement center, or individual deployment sites. All military, civilian, and contractor personnel destined for joint or multinational positions attend training at the activity. The supported geographic combatant commander may waive the requirement for an individual to train at the center on a by-exception basis. (See Department of the Army Pamphlet (DA Pam) 600-81 for more information.)

5-36. Integration and training occur once the JTF is formed and begins its work. The integration occurs in two environments: when the augmentee arrives at the JTF location and when the augmentee is assigned to a staff section or other staff element. Integrating the JTF staff proves challenging whether during the initial organizational phase or the training and replacement of individuals and capabilities for a mission of long duration. The JTF headquarters commandant, normally the corps headquarters battalion commander, is usually identified as the integration point of contact. This commandant controls reception, administrative processing and life support resources—billeting, messing, transportation—that are the initial concern. With oversight by the corps chief of staff, the JTF staff establishes procedures for reception, initial orientation, personnel accountability, and strength reporting, training, and security.

AUGMENTATION TO THE CORPS HEADQUARTERS

5-37. The corps commander and staff may not have the required expertise to fill all JTF positions based on the mission. Some are readily available from the combatant commander and Army sources, but others require lead time and must be formally requested, especially one-of-a-kind national-level assets such as the joint communications support element. Paragraphs 5-42 through 5-52 discuss joint organization augmentations to the corps headquarters. JP 3-33 addresses this subject in some detail. Figure 5-3 portrays some of the various enablers available to augment the corps headquarters forming it into a joint headquarters.

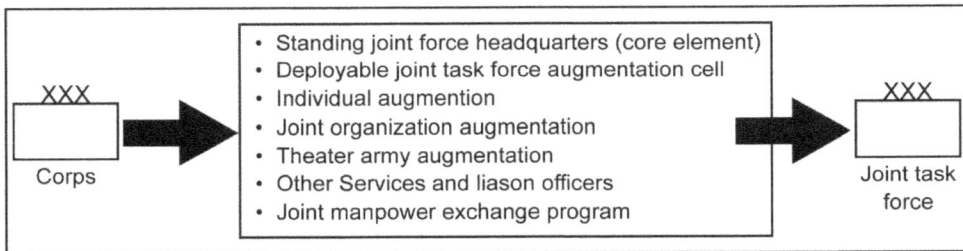

XXX

Corps

- Standing joint force headquarters (core element)
- Deployable joint task force augmentation cell
- Individual augmention
- Joint organization augmentation
- Theater army augmentation
- Other Services and liason officers
- Joint manpower exchange program

XXX

Joint task force

Figure 5-3. Augmenting the corps staff

STANDING JOINT FORCE HEADQUARTERS (CORE ELEMENT)

5-38. The standing joint force headquarters (core element) (SJFHQ(CE)) is a full-time, cross-functional command and control element on the combatant commander's staff. Commanded by a general or flag officer, it fully integrates into the geographic combatant commander's planning and operational activities and stands ready to conduct deliberate or crisis action planning in support of current or future operations. Augmenting a corps headquarters with an SJFHQ(CE) provides a minimum joint capability. The geographic combatant commander directs its internal organization that is usually arranged into functional teams. A colonel commands each team. This officer often has joint experience in planning, operations, information superiority, knowledge management, and logistics. Depending on the situation, additional groups are formed, generally to provide capabilities not normally found in a corps headquarters. See figure 5-4 (page 5-9) for an example of SJFHQ(CE) support to a corps headquarters.

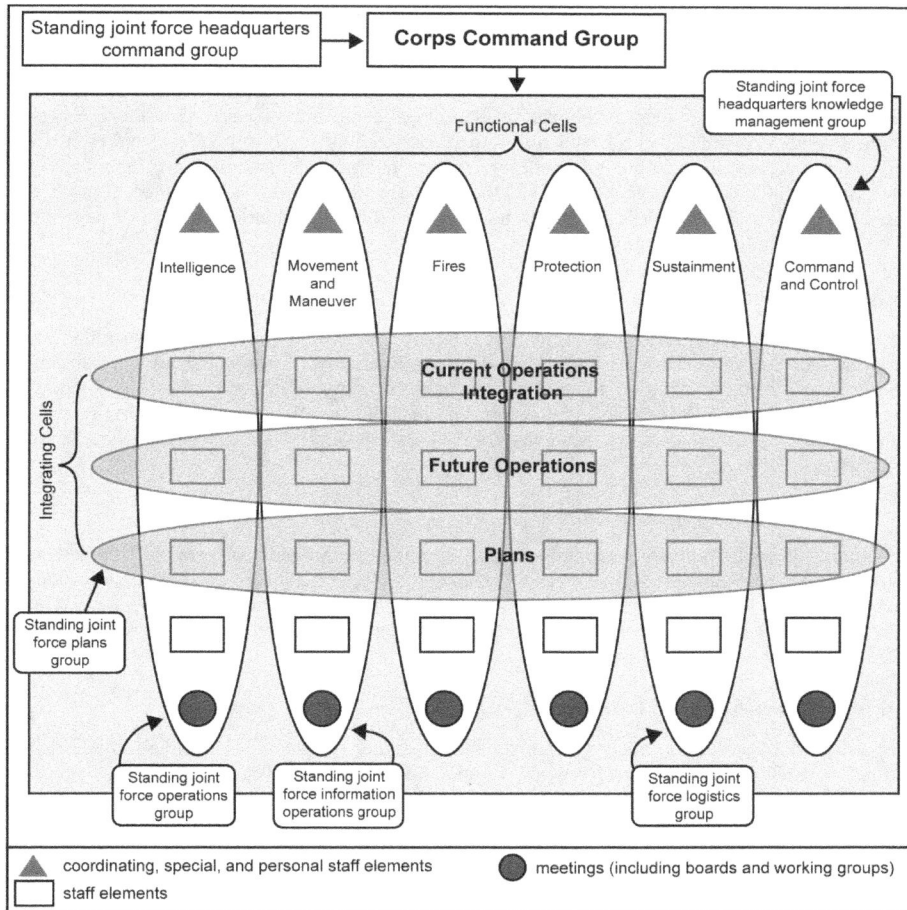

Figure 5-4. Standing joint force headquarters support to a corps headquarters example

5-39. The SJFHQ(CE) is used in three modes: the core of a JTF headquarters for the geographic combatant commander, a specialized cell within the geographic combatant commander, and a joint augmentation for a Service component as the JTF headquarters. A corps headquarters designated as a JTF headquarters will most likely see the latter option. The SJFHQ(CE)'s purpose is to jump-start the joint planning process with a trained, well-equipped plug. When serving as the core of a JTF headquarters, the corps can expect to receive a SJFHQ(CE). Depending on the level of activity in the combatant commander's area of operations, the SJFHQ(CE) may stay for the length of the mission or redeploy—in whole or in part—before mission accomplishment.

DEPLOYABLE JOINT TASK FORCE AUGMENTATION CELL

5-40. The deployable JTF augmentation cell is similar to a SJFHQ(CE), but it is not a standing organization. It is a tailored pool of augmentees hand-picked for their expertise and trained in crisis action procedures. Commanders use the deployable JTF augmentation cell like a standing joint force headquarters. This cell may be phased out as the SJFHQ(CE) concept matures and combatant commanders develop JMDs.

INDIVIDUAL AUGMENTATION

5-41. Every forming JTF headquarters receives individual augmentees. They can be Army personnel identified and requested by the corps commander and principal staff or plugs to fill slots in the JMD. Identifying suitable individual augmentees is one of the earliest tasks the corps staff must perform. Determining the number, skill set, type, and availability of augmentees occurs while the JTF is still in the forming stages. This enables the headquarters to request military, Army civilians, representatives from other government agencies, and contractors. Frequently, individual augmentees are functional experts to chair, provide guidance, or serve on meetings (to include working groups and boards), centers, cells, and planning teams associated with joint operations. As with other augmentation, individual augmentees may not remain until mission accomplishment.

JOINT ORGANIZATION AUGMENTATION

5-42. Several joint organizations exist to augment the joint community in the execution of military operations. As the joint force provider, United States Joint Forces Command arranges for these joint enabling capabilities. In addition to deployment support, many of these entities can support training exercises as resources allow. Some joint enabling capability entities are self-supporting, while others require support from the supported headquarters. The entities in JP 3-33 are valuable assets to the JTF headquarters.

Joint Communications Support Element

5-43. The joint communications support element is a joint command that provides rapidly deployable communications augmentation. It consists of a headquarters support squadron and communications support detachment, three active squadrons, two Air National Guard squadrons, and one Army Reserve squadron. As a low-density, high-value asset, the joint communications support element may not remain in support of a JTF headquarters for the duration of its mission.

National Intelligence Support Team

5-44. A national intelligence support team provides a rapidly deployable and mission-tailored national intelligence reachback capability to provide a national-level, deployable, all-source intelligence team to meet a JTF's intelligence requirements. National intelligence support teams are nationally sourced and composed of expert intelligence and communications analysts, communicators, and managers from the Defense Intelligence Agency, Central Intelligence Agency, National Security Agency, National Geospatial-Intelligence Agency (formerly the National Imagery and Mapping Agency), and other agencies.

Defense Threat Reduction Agency

5-45. The Defense Threat Reduction Agency is a Department of Defense agency providing subject matter expert augmentees to counter weapons of mass destruction (chemical, biological, radiological, nuclear, and high-yield explosives). This agency provides capabilities to reduce, eliminate, and counter the threat, and mitigate its effects. It supports combatant commanders and JTF staffs with specialists on developing necessary offensive and defensive tools, and equipping Soldiers for chemical or biological attacks.

Joint Information Operations Warfare Command

5-46. The United States Strategic Command's Joint Information Operations Warfare Command augments planning, coordinating, and executing efforts for the joint information operations community. It rapidly deploys information operations planning teams to deliver tailored, highly skilled support and sophisticated models and simulations to joint commanders and JTF headquarters. Its core capabilities include command and control warfare, computer network operations, psychological operations, military deception, and operations security. These capabilities work with specified supporting and related capabilities to influence, disrupt, corrupt, or usurp adversarial human and automated decisionmaking while protecting friendly operations and organizations. A reachback capability enables it to respond to emerging situations within a joint operations area.

Joint Personnel Recovery Agency

5-47. The Joint Personnel Recovery Agency is a United States Joint Forces Command subordinate serving as the principle Department of Defense agency for coordinating and executing personnel recovery. With regards to JTF operations, this agency augments personnel recovery efforts in four areas: guidance, education and training, operational support (including exercises and deployments), and lessons learned and research and development.

Joint Public Affairs Support Element

5-48. United States Joint Forces Command's joint public affairs support element augments the geographic combatant commander and JFC with a rapidly deployable, trained, equipped, and expert team with knowledge in joint public affairs, media operations and both Service and joint policies. Training teams for the joint public affairs support element provide a standing, rapidly-deployable, turn-key joint public affairs capability to support various operational requirements. Each training team forms the core of a scalable public affairs response capability, a ready, mission-tailored force package to support exercises and to deploy in support of the combatant commands for operations and contingencies.

Defense Logistics Agency

5-49. The Defense Logistics Agency supports the JTF using various capabilities, to include Defense Logistics Agency contingency support teams and other experts imbedded physically or virtually with the JTF. These teams provide liaison officers and functional experts with logistic planning experience, logistic surge, and sustainment expertise to the agency. Team capabilities include logistics assistance teams to address supply management issues, disposal remediation teams to manage disposal of hazardous waste, distribution operations teams to provide expertise in distribution management, and fuel support teams to serve as a liaison in bulk fuel operations. (For more information, see the Defense Logistics Agency Web site, listed in the References.)

Deployable Joint Command and Control System

5-50. The deployable joint command and control system is a command and control element providing the geographic combatant commander and subordinate JTF headquarters with a full range of interoperable, robust, standardized, and scalable systems and tools for planning, executing, and assessing joint operations. This system provides the integrated hardware and software suites that allow commanders to exercise command and control over widely dispersed forces using multiple data sources and communications alternatives.

Joint Fires Integration and Interoperability Team

5-51. The Joint Fires Integration and Interoperability Team provides rapidly deployable battlefield assessment teams to augment large-scale training exercises and operational deployments to gather data on the planning, preparation, and execution of joint fires integration. When deployed, they focus on joint intelligence, surveillance, and reconnaissance, joint air-to-ground fires integration with maneuver, command and control, and combat identification. Each battlefield assessment team aims to produce effective target acquisition, command and control, and interoperable firing systems to reduce fratricide and collateral damage. (For more information, see the Joint Fires Integration and Interoperability Team Web site, listed in the References.)

Joint Systems Integration Center

5-52. United States Joint Forces Command's Joint Systems Integration Center brings together operational and technical expertise, technology, state-of-the-art facilities, and repeatable scientific methodology to augment joint command and control capabilities, and solve joint interoperability problems, focusing at the JTF level. It ensures identified capabilities are interoperable from the geographic combatant commander level down through the JTF and its subordinates. The end result is a recommendation that will lead to improved interoperable capabilities. Onsite assistance and reachback capability support the JTF. (For more information, see the Joint Systems Integration Center Web site, listed in the References.)

THEATER ARMY AUGMENTATION

5-53. Wherever deployed, the corps headquarters serving as a JTF headquarters falls under the administrative control of a theater army. Depending on the type and duration of the mission, the theater army may provide individual or unit augmentation from the following Army organizations to enable the JTF headquarters to accomplish its mission. These include—

- Theater sustainment command.
- Signal command (theater).
- Military intelligence brigade.
- Civil affairs brigade.
- Medical command (deployment support).
- Theater aviation command.

5-54. Other potential augmenting organizations include a battlefield coordination detachment, Army air and missile defense command, and other functional organizations. These organizations often deal with contracting; chemical, biological, radiological, nuclear, and high-yield explosives (CBRNE); engineering; military police; information tasks; aviation; and space support.

OTHER SERVICES AND LIAISON OFFICERS

5-55. Liaison often forms the glue that holds joint, coalition, and multinational operations together. The JTF headquarters provides trained and knowledgeable liaison teams. Both the composition of the teams and the number required is maintained throughout the mission. Liaison detachments have four basic functions: monitor, coordinate, advise, and assist. Each team sent to superior, subordinate, adjacent, or supporting organizations has enough staff to provide 24-hour coverage and sufficient rank structure to gain access to the appropriate level at the receiving command. Each liaison team received from a sending unit is integrated into the staff and accommodated with work space, communications, protection, life support, and sufficient access to the commander and appropriate staff to accomplish its mission. Each corps liaison team has a magnified challenge when it sets up at distant locations. Often the teams have limited transportation assets to receive guidance, supplies, mail, and morale support.

5-56. Current doctrine identifies the chief of staff as the central point of contact for liaison operations. However, the sheer number of liaison teams in joint operations with a large military, intergovernmental, and multinational force may make that impractical. An alternative involves establishing a liaison office in the main CP associated with the movement and maneuver cell. This office contains a director and a small staff to control liaison operations, both those sent from and those received by the corps headquarters. An individual of sufficient rank directs and staffs the office to control the operation and has ready access to corps senior leaders to facilitate information exchange. Liaison office functions include maintaining liaison rosters, managing status reporting and accountability, serving as an information clearinghouse, and providing communications systems. The liaison office also acts as the single point of contact for life support and other liaison support operations, such as protection, medical assistance, and security.

5-57. Digital liaison detachments augment corps headquarters, especially when serving as a JTF or joint force land component command. Each detachment of 30 personnel can establish a liaison with a multinational division or higher headquarters and provide digital connectivity with ABCS for maneuver, fires, intelligence, sustainment, and air and missile defense. Alternatively, these detachments can be split into two or more teams to provide connectivity to two or more brigade-sized elements.

5-58. Adequate liaison is most important to a corps headquarters transitioning to a joint headquarters in those critical situations that will determine success or failure. Examples include situations during predeployment academic and training exercises; during reception, staging, onward movement, and integration; before new equipment training; and during the first few weeks as a joint headquarters.

JOINT MANPOWER EXCHANGE PROGRAM

5-59. The joint manpower exchange program places qualified military personnel in billets outside their Service. When mature, this program will ensure that each geographic combatant commander and many of their Service components commands have officers from another Service embedded in their staffs. Such a program provides in-house liaison and staffs knowledgeable of both their specialty in their parent Service and in the operations of the organization in which they serve.

OTHER AUGMENTATION OR COLLABORATIVE CAPABILITES

5-60. Capabilities can be augmented or collaborative. The joint enabling capabilities are a type of augmentation. The interagency enabling capabilities; linguistic, interpreter, and cultural support; and multinational enabling capabilities do not augment. Rather, they are often collaborators focusing on the same objectives. They often join the corps headquarters serving as a JTF headquarters after it has arrived in the joint operational area.

INTERAGENCY ENABLING CAPABILITIES

5-61. Modern operational environments contain more than just military organizations, especially if the JTF is engaged in stability or civil support activities. The Army corps headquarters as a JTF headquarters can expect to interact with and perhaps receive and provide direct support to many other government agencies. These include the American embassy country team, the Department of State, Department of Homeland Security, Department of Agriculture, Department of Commerce, and Department of Transportation. Likewise, independent agencies of the federal government often are involved in military operations, including the Central Intelligence Agency, the Environmental Protection Agency, National Aeronautics and Space Administration, and especially the United States Agency for International Development. The list expands considerably when including the state and local government entities.

5-62. Overseas, the corps as a JTF headquarters comes in contact with representatives of international organizations, such as the United Nations, International Committee of the Red Cross, International Red Cross and Red Crescent Movement, and the hundreds of nongovernmental organizations. While many of these organizations try to remain neutral by not associating directly with the JTF, many interact with it as they pursue their specialized missions. Host-nation governments, whether they welcome the U.S. presence or not, will interact with the JTF headquarters and will have to be accommodated. (See JP 3-08 for further information on interagency, intergovernmental, and nongovernmental coordination.)

5-63. Interagency capabilities provide skills lacking in sufficient quantities to support mission accomplishment. Inadequate staffing and training of required interagency personnel can inhibit this process. The interagency representative provided by non-Department of Defense activities of the United States Government may come with a skill set not fully ready for the mission at hand. Early in the forming, orientation, and training of JTF members, JTF leadership must identify knowledge shortfalls and provide the training to fill the knowledge gaps. This orientation and training may involve providing equipment to facilitate access to the common operational picture and JTF databases.

LINGUIST, INTERPRETER, AND CULTURAL SUPPORT

5-64. Linguists and interpreters (to include translators) often are critical to JTF operations. Interpreters transfer the meaning of one spoken language into another spoken language. Translators render the meaning of one written language into another written language. Foreign deployments often require language proficiency—especially for local dialects—that require language skills beyond those typically resident in the geographic combatant commander or JTF staffs. Linguists can provide valuable training to the JTF staff and key personnel for specific joint operations area. Early in the planning and forming stages of the JTF's lifecycle, staff identify and resource the requirements for linguists, interpreters and translators. Integrating interpreters and translators into the JTF occurs as soon as possible to obtain security clearance and finalize contractual agreements. In-place procedures are required to identify the interpreter chain of command and the scope of interpreter duties. The corps headquarters develops procedures to vet interpreters and translators to uncover biases. The staff considers the possibility that they may not have the best interests of the JTF foremost in their minds. Other positions may be sourced through vetted, contracted interpreters.

5-65. Cultural support goes beyond the use of local interpreters and local hires to perform life support tasks. Inserting U.S. forces into a different culture creates a dynamic to address. Each JTF requires cultural intelligence to successfully deal with the local population. Knowledge of the local culture in terms of the social, political, economic, and demographic factors contributes to understanding a people or a nation's history, institutions, psychology, religious beliefs, and behaviors. Increasing cultural awareness using Department of State regional experts, special operations forces with regional expertise, or civilian academics with language and cultural understanding facilitates the planning, preparation, and execution of military operations. Additionally, foreign area officers from the security cooperation division of the theater army headquarters may serve as political or cultural advisors to the corps or subordinate commanders.

MULTINATIONAL ENABLING CAPABILITIES

5-66. When deployed outside the continental United States, the U.S. military rarely operates alone. Whether a part of a formal treaty organization such as the North Atlantic Alliance or as an ad hoc coalition like that for Operation Iraqi Freedom, the JTF headquarters multinational participation complicates normal unilateral organization, planning, and operations. Operating in a multinational environment presents challenges as well as opportunities. A JTF operating with multinational forces follows multinational doctrine and procedures if the United States has ratified that doctrine and those procedures. A lead nation uses its national doctrine for operations. When no multinational doctrine exists, the JTF follows U.S. joint doctrine where possible. For doctrine and procedures not ratified by the United States, commanders evaluate and follow the multinational command's doctrine and procedures, where applicable and consistent with U.S. law, regulations, and doctrine.

5-67. Command relationships, interoperability, combined planning, information and intelligence sharing, communications systems, and logistic support complicate JTF operations. However, the opportunity to take advantage of embedded linguistic and cultural competence, regional and local intergovernmental contacts, and military expertise of multinational partners makes the effort worthwhile. Most often, the United States serves as the lead nation with the largest forces on the ground, but effective exchange of liaison teams and combined preparation, planning, and execution facilitates JTF operations.

5-68. The JTF headquarters with a multinational component facilitates integration. It prompts thoroughly understanding national restrictions on operations, organization of the force, sustainment requirements, early identification of command relationships, and responsibilities and expectations. Clarity of these considerations at the start of operations enables the JTF headquarters to get the most from its multinational forces. See JP 3-16 and the *ABCA Coalition Operations Handbook* for additional information on multinational enabling capabilities.

JOINT TASK FORCE HEADQUARTERS EQUIPPING CAPABILITIES

5-69. Equipping a corps headquarters as it transitions to a JTF headquarters goes hand-in-hand with force generation and augmentation. No single Service possesses all the command and control information systems and equipment required by a JTF headquarters. A Service headquarters allocated to the geographic combatant commander is a JTF headquarters so the former can provide a core element that has engaged in preparation and certification, to include identifying the required equipment and accompanying shortfalls. The joint enabling capabilities can supply, either permanently or on a temporary basis, shortfalls or shortages in equipment. Most equipment necessary to give the JTF full capabilities is acquired from other sources to include commercially. A start on this activity is a detailed joint mission-essential equipment list, most of which will be electronic communications equipment.

5-70. The joint mission-essential equipment list consists of required equipment and capabilities to build on the base. Equipment can be added in blocks, each block to incrementally plus up the JTF headquarters to meet the specific projected or mission demands. For a corps headquarters, the foundation block consists of its organic Service-provided capabilities, followed by equipment received from the theater army and the theater signal command or brigade. Usually these capabilities—internal systems—consist of computers with software applications. Such capabilities consist of operating systems, office applications and security, associated peripherals, and servers. To fill the gaps, units identify any continuing shortfall between needs and capability, refine a joint mission-essential equipment list, and submit a request through joint or Army channels.

5-71. For the Army-based JTF headquarters and associated Army units, equipping a transitioning corps headquarters involves the theater army. The equipment available in theater comes from three sources: theater property, Army pre-positioned stocks, and theater sustainment stocks. The theater army, geographic combatant commander, and the joint enablers sometimes cannot provide equipment. Some equipment is not commercially available. In those cases, units can source up the chain of command to the joint staff to process an operational needs statement. Often, needed equipment includes technical systems related to worldwide communications, additional bandwidth, and video, voice, and data services from military and commercial sources. Deployable joint enablers such as the joint communications support element and the deployable joint command and control system element are the first line of support for outside assistance. (Appendix C provides additional information on communication support.)

JOINT LAND OPERATIONS

5-72. The Army corps headquarters can serve as joint force land component command headquarters. Land operations support the JFC's operation or campaign objectives or support other components of the joint force. Joint land operations require synchronizing and integrating the instruments of national power to achieve strategic and operational objectives. These forces conducting joint land operations include the Army, Marine, and Naval forces operating on or from land to accomplish missions and tasks. Normally, joint land operations also involve multinational land forces. Joint land operations specifically include land control operations. They employ land forces, supported by naval and air forces (as appropriate), to accomplish military objectives in vital areas of the operational area.

JOINT FORCE LAND COMPONENT ORGANIZATION

5-73. When organizing joint forces, simplicity and clarity are critical: by providing the joint force land component a single commander for joint land operations, the JFC can enhance synchronization of operations not only between U.S. ground components, but with multinational land forces as well. Forming a joint force land component builds unity of effort, an integrated staff, a single voice for land forces and land control operations, one single concept and focus of effort for land control operations, and a synchronized and integrated land force planning and execution. The disadvantages are that joint force land component normally retains Service component responsibilities to the JFC (requires split focus of the staff), it challenges integrating staffs, it requires more lead-time to establish headquarters before execution, and it lacks the ability to resource the staffs. See JP 3-31 for additional doctrinal guidance on establishing the joint force land component.

CORPS HEADQUARTERS AS A JOINT FORCE LAND COMPONENT HEADQUARTERS

5-74. Normally, a Service headquarters provides the joint force land component commander (JFLCC). If possible, the joint force land component may have a separate Army forces or Marine forces commander and headquarters responsible for the administrative control of the respective Services in the land component. If not, the JFLCC continues to be responsible for Service component functions. This latter arrangement has the potential to over task the JFLCC's staff while performing its dual role. It may be advantageous for the JFLCC to delegate as many of duties related to Service component as practical to a subordinate Service force headquarters.

5-75. Within the joint force land component headquarters, the corps commander, deputy commander, chief of staff, and key members of the staff fully integrate with representation from the forces and capabilities made available to the JFLCC. The staff includes the manpower and personnel directorate of a joint staff (J-1) through the communications system directorate of a joint staff (J-6). The corps commander when designated as the JFLCC provides the core elements of the staff to assist in planning, coordinating, and executing functional land component operations. See JP 3-31 and Chairman of the Joint Chiefs of Staff Instruction (CJCSI) 1301.01C.

5-76. Combatant commanders follow appropriate guidance for their assignments. An annually updated "Forces for Unified Commands" memorandum from the Secretary of Defense assigns forces to combatant commanders. During crisis action planning, forces allocated to combatant commanders may differ from those apportioned for contingency planning. For more information about how Department of Defense assigns forces, see CJCSI 3100.01B and Chairman of the Joint Chiefs of Staff Notice (CJCSN) 3500.01.

JOINT FORCE LAND COMPONENT COMMAND HEADQUARTERS

5-77. The headquarters is organized according to the JFC's implementing directive. This document establishes the roles and responsibilities of the JFLCC and designates the mission and forces. Normally, the staff will be built around the corps staff and augmented with members of the other Service component or forces as described in paragraphs 5-37 through 5-59. The JFLCC's staff allocates key staff billets so that all Services are appropriately represented and shared equitably in staffing tasks. Ideally, the deputy JFLCC or chief of staff comes from a different Service. Replicating this construct throughout the staff leadership ensures all leaders understand the distinct capabilities of each Service to optimize employment of the forces. See JP 3-31 for a depiction of a notional joint force land component headquarters organization.

JOINT FORCE LAND COMPONENT COMMAND RESPONSIBILITIES AND ROLES

5-78. The corps headquarters serving as a joint force land component headquarters exercises responsibilities under the authority of the combatant commander. The commander of the joint force land component has responsibilities that include, but are not limited to, the following:

- Advise the JFC on the proper employment of forces and capabilities.
- Develop joint plans and orders in support of the JFC's concept of operations and optimize the operations of task-organized land forces.
- Execute and assess land control operations.
- Coordinate the planning and execution of joint land operations with the other components and supporting agencies.
- Synchronize and integrate all aspects of combat power in support of land operations.
- Designate the target priorities, effects, and timing for joint land operations.
- Establish a personnel recovery element to account for and report the status of isolated personnel and to coordinate and control land component personnel recovery events.
- Provide mutual support to other components by conducting operations within the operational area.
- Coordinate with other functional and Service components in support of accomplishment of JFC's objectives.
- Provide an assistant or deputy to the area air defense commander for land-based joint theater air and missile defense operations as determined by the JFC.
- Support the JFC's information operations by developing the information operations requirements that support land control operations and synchronize land force information operations assets when directed.
- Integrate the joint force land component's command and control system and resources into the theater's command and control system.
- Integrate special operations as required into overall land operations.
- Perform joint security functions.
- Supervise detainee operations.
- Establish standing operating procedures and other directives based on the JFC's guidance.
- Provide inputs to the JFC-approved joint area air defense plan and the airspace control plan.
- Assess and as necessary restore or reconstruct civilian infrastructure.

Appendix A

Sustainment

As a part of its transformation, the Army is changing how it sustains its forces. The concept of centralized control with decentralized execution is a new one for logistics but not new to the Army. This appendix discusses the historical background for transformation, transformation, functions of sustainment, corps sustainment, and corps support requirements.

BACKGROUND

A-1. In World War II, Army doctrine established that the commander of the communications zone was co-equal to the commander of the Army Air Forces and the commander of Army ground forces in the combat zone. While most Service forces were retained at theater-level, selected Service units provided direct support to field armies. After World War II, this was reversed with most sustainment units subordinated to maneuver commanders at every echelon.

TRANSFORMATION

A-2. Transformation in Army sustainment doctrine in support of full spectrum operations at the operational and tactical levels is provided by highly trained modular sustainment forces, integrated and synchronized with the operational plan. The goals of sustainment transformation are to develop a sustainment infrastructure—

- More responsive to the needs of an expeditionary and joint capable force.
- Free of redundancy.
- Streamlined (by reducing unnecessary layers).

A-3. Responding to these imperatives, sustainment leaders have designed a capability that leverages emerging technologies and links support to supported organizations and the Army-to-joint organizations from the continental United States to areas of operations (AOs) and within those AOs. The design of the capability accounts for the explicit guidance from Headquarters, Department of the Army that makes no Reserve Component sustainment units available within the first 30 days of an operation. The design also clarifies that the sustainment force will be as capable as the current sustainment force after transformation.

SUSTAINMENT FUNCTIONS

A-4. Sustainment is the provision of personnel, logistic, and other support required to maintain and prolong operations until mission accomplishment. The *sustainment warfighting function* is the related tasks and systems that provide support and services to ensure freedom of action, extend operational reach, and prolong endurance (Field Manual (FM) 3-0). It includes the functions of—

- **Personnel services.** These services include human resources support, financial management, legal support, religious support, and band support functions related to Soldiers' welfare, readiness, and quality of life.
- **Logistics.** Logistics is the military art and science of carrying out the movement and maintenance of forces. Logistics includes maintenance, transportation, supply, field services, distribution, operational contract support, and general engineering support.
- **Health service support.** This includes all support and services performed, provided, and arranged by the Army medical department to promote, conserve, or restore the mental and physical well-being of Army personnel and, as directed, personnel in other Services, agencies, and organizations. This includes casualty care, medical evacuation, and medical logistics.

A-5. The force structure required to execute effective and efficient sustainment operations is never static. As the theater matures and operational requirements change, the modular sustainment structure also changes in anticipation of operational requirements.

A-6. While basic sustainment functions do not change as the corps headquarters assumes the missions, staffing does change. As discussed in chapter 5, transitioning from an Army corps headquarters to a joint headquarters requires significant augmentation, including Army augmentation to the corps headquarters. The level and type of augmentation depends on the joint commander's intent, mission, and concept of operations.

CORPS SUSTAINMENT

A-7. Normally, modular sustainment forces are assigned or under operational control (OPCON) to the TSC for the theater army with support provided at every echelon of command: theater army, corps, division, and brigade. Integral to the success of modular sustainment is its ability to leverage and synchronize support from joint and national or strategic partners. These partners can include the United States Transportation Command, the Defense Logistics Agency, Air Mobility Command, General Services Administration, and United States Army Materiel Command. The corps main command post (CP) sustainment cell monitors each of the four functions of sustainment with a different staff principal taking the lead:

- Assistant chief of staff, personnel (G-1) leads the function of personnel.
- Assistant chief of staff, logistics (G-4) leads logistics.
- Assistant chief of staff, financial management (G-8) leads the function of resource management and coordinates with the appropriate level financial management support officer—TSC or expeditionary sustainment command (ESC)—for finance operations capability.
- Surgeon leads Army Health System support.

A-8. Normally the G-4 serves as both the chief of sustainment and the logistics section chief, but the corps commanding general may also designate a chief of sustainment. Requisitions and requests for assistance are initiated at the user level and processed in the TSC support chain. The main CP sustainment cell monitors the process. A sustainment organization is assigned, OPCON, or in support of units at every echelon. The TSC and its forward echelon ESC provide the support. Whether provided from a brigade support battalion, a sustainment brigade, a TSC or ESC, or the national or strategic-level, the corps benefits from the transformed Army's centralized execution model.

A-9. As depicted in figure A-1, the corps and its assigned, attached, OPCON, or tactical control (TACON) units receive their support from a number of modular sustainment units. Medical and other sustainment comes from the theater army or the continental United States force generation base and the brigade combat team from its organic brigade support battalion.

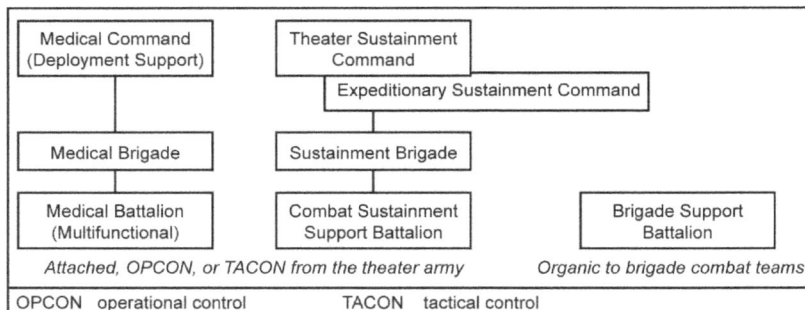

Figure A-1. Major components of the modular force sustainment structure

MEDICAL COMMAND (DEPLOYMENT SUPPORT)

A-10. The medical command (deployment support) is the senior medical command within a theater. It provides the requisite command and control necessary to deliver timely and responsive Army Health System support for deployed forces. The medical command (deployment support) is a dedicated, regionally focused command with a basis of allocation of one per theater. This command provides subordinate medical organizations that operate under the medical brigade and/or medical battalion (multifunctional). It also provides medical augmentation as forward surgical teams or other augmentation required by supported units.

Medical Brigade

A-11. The medical brigade is designed to provide command and control to medical units supporting the corps and other units in the theater of operations. The design and flexibility of the medical brigade enables the supported commander to meet the medical support requirements of early-entry forces. As the supported commander's forces grow in size and complexity, the medical brigade can deploy additional modules that build upon one another to support operations. The medical brigade also provides the commander the appropriate medical command and control to continue to build medical force capabilities by integrating Army, joint, and multinational medical forces to ensure the ability to identify and counter health threats in the AO. The medical brigade is assigned to the medical command (deployment support) and may be attached or OPCON to a corps or division headquarters depending on the situation.

Medical Battalion (Multifunctional)

A-12. The medical battalion (multifunctional) exercises flexible command and control and provides administrative assistance, medical logistics, and technical supervision of assigned and attached functional medical organizations.

Medical Augmentation

A-13. Based on the situation, the expected duration of the operation, and any unique requirements (such as being parachute qualified), teams to address anticipated requirements augment medical formations. Augmentation is either a team or an individual augmentee. Teams such as plans and operations, clinical services, preventative medicine, mental health, veterinary services, medical logistics, and optometry are available to meet the need. Individual professional fillers also augment units. The professional filler system provides augmentees with low-density, high-demand medical qualifications.

EXPEDITIONARY SUSTAINMENT COMMAND

A-14. The ESC deploys to exercise command and control when multiple sustainment brigades are employed or when a forward command presence is required. The ESC provides operational reach and improved span of control for the TSC. The ESC functions as an extension of the TSC rather than another echelon of command. The ESC plans and executes sustainment; distribution; theater opening; and reception, staging, and onward movement for Army forces. It may serve as the basis for an expeditionary joint command when directed by the geographic combatant commander or designated coalition or joint task force commander.

Sustainment Brigade

A-15. The sustainment brigade is a subordinate command of the TSC. It is a multifunctional sustainment organization, tailored and task-organized according to the situation. It conducts sustainment operations within its specified AO. Sustainment brigades consolidate selected functions previously performed by corps and division support commands and area support groups into a single operational echelon. These brigades exercise command and control of theater opening, theater distribution, and sustainment operations. Their core competency is command and control of sustainment operations. Normally sustainment brigades are used in a supporting to supported relationship. See FMI 4-93.2.

A-16. Sustainment brigades are an integral component of the joint and Army battlefield communications network. They use satellite and network-based communications that enable command and control, have visibility of the distribution system, and identify support requirements. However, not all sustainment brigades have an organic network signal element. Those without this capability rely on theater-level signal network capability for command and control network integration.

Combat Sustainment Support Battalion

A-17. The combat sustainment support battalion is the building block on which TSC sustainment capabilities are developed. Typically attached to a sustainment brigade, the combat sustainment support battalion is tailored to meet specific mission requirements. Attached capabilities drawn from the force pool can include transportation, maintenance, ammunition, supply, mortuary affairs, airdrop, field services, water, and petroleum.

A-18. Employed on an area basis, the combat sustainment support battalion plans, coordinates, synchronizes, monitors, and controls sustainment operations (less medical) within a specified AO. It supports units in or passing through its geographic area.

BRIGADE SUPPORT BATTALION

A-19. Brigade support battalions are organic to brigade combat teams and multifunctional support brigades (except for the battlefield surveillance brigade). The battlefield surveillance brigade is supported by a brigade support company. Their capabilities are tailored to the specific type of brigade they support. For example, the brigade support battalion of a heavy brigade combat team has more fuel distribution and maintenance capabilities than a fires brigade. Brigade support battalion capabilities include supply, maintenance, motor transport, and medical support. They plan, coordinate, synchronize, and execute logistic operations in support of brigade operations.

CORPS SUSTAINMENT REQUIREMENTS

A-20. Effective sustainment from a corps commander's perspective ensures freedom of action, extends operational reach, and prolongs endurance. Effective sustainment provides responsive and agile support at the right place, at the right time, and in the right quantity. Regardless of the perspective, successful sustainment operations depend on a commander understanding requirements, capabilities, priorities, and the operational environment.

A-21. In most instances, the TSC and its subordinate organizations maintain a supporting to supported relationship with the corps. The actual command and control relationship is usually specified by an operation order. Collaborative planning and coordination among the TSC, ESC, sustainment brigade, and the corps main CP sustainment cell provides the situational understanding necessary for synchronizing and integrating sustainment operations with the corps battle rhythm. Paragraphs A-22 through A-64 describe the sustainment warfighting function in the context of corps operations.

PERSONNEL SERVICES

A-22. Personnel services include human resources support, financial management support, legal support, religious support, and band support functions related to Soldiers' welfare, readiness, and quality of life.

Human Resources Support

A-23. Human resources support to corps units is provided by TSC sustainment brigades that provide support on an area basis within their specified AO. Realigned human resources capabilities result from the emergence of the sustainment warfighting function, clarifications of definitions, and the loss of traditional human resources command and control structure above company level in the modular force structure.

A-24. Depending on the function a sustainment brigade performs—theater opening, theater distribution, or sustainment—it exercises command and control of human resources companies, military mail terminal teams, theater gateway personnel accountability teams, or a combination of these. A human resources sustainment center, functioning as a staff element of the TSC, provides technical guidance and ensures execution of human resources support as defined by the policies and priorities established by the Army Service component command and G-1. A sustainment brigade human resources branch plans, coordinates, integrates, and manages human resources support within the sustainment brigade's specified AO.

A-25. Human resources unit capabilities include personnel accounting, casualty operations, and postal operations. Each capability is fully integrated and synchronized with all other facets of the sustainment function. This integration effectively and efficiently sustains units in or passing through the sustainment brigade's specified AO. For more detailed information on human resources support, see FM 1-0 and FMI 4-93.2.

Financial Management Support

A-26. Financial management operations support to corps units is provided by TSC sustainment brigades that provide support on an area basis within their specified AO. A financial management center that functions as a staff element of the TSC provides technical oversight of finance operations.

A-27. Subordinate financial management company and detachment capabilities include determining currency requirements and replenishment, receiving collections, and making payments on certified vouchers (commercial vendor services and payments). Other capabilities include conducting enemy prisoner of war and civilian internee support; safeguarding funds and protecting funds from fraud, waste, and abuse; and providing pay support. See FM 1-06 and FMI 4-93.2.

Legal Support

A-28. Corps offices of the staff judge advocate provide legal support to strategic-level planning of operations. They further support the efforts to advise commanders of division offices of the staff judge advocate and brigade legal sections. Corps offices of the staff judge advocate also provide analysis and advice regarding lower-echelon legal actions that require broader oversight due to law, regulation, or policy. When deployed, irregular operational efforts may require direct contact between brigade and corps legal personnel. Corps offices of the staff judge advocate maintain the capability to analyze specific brigade mission requirements. As with a better-resourced organization, reporting requirements flow upward, but the general burden of support flows from the corps office of the staff judge advocate to the division office of the staff judge advocate to the brigade legal section. FM 1-04 discusses legal support.

A-29. The staff judge advocate advises the commander on legal obligations concerning the local population, detained and displaced persons, and on other matters. Those other matters can include combat contingency contracting, fiscal law, processing claims arising in an operational environment, and environmental law. Legal personnel frequently serve in support of stabilization efforts led by other entities, such as a civil affairs section or other United States Government agency.

A-30. Judge advocates serve at all levels in today's operational environment and advise commanders on various operational legal issues, including the law of war, rules of engagement, lethal and nonlethal targeting, treatment of detainees and noncombatants, fiscal law, foreign claims, contingency contracting, the conduct of investigations, and military justice. They also serve as staff officers and on boards, centers, and cells, where they fully participate in the planning process.

A-31. Legal support in today's operational environment applies to all warfighting functions. Typically, staff judge advocate personnel assist in command and control, sustainment, and personnel functional areas. However, judge advocates and paralegal Soldiers assist in planning and operations that bridge all areas of full spectrum operations.

A-32. Legal support to personnel services support includes the operation of each command's claims program and supervision of the area claims office or claims processing office designated by the United States Army Claims Service. It also includes providing personal civil legal services to Soldiers, their family members, and other eligible personnel.

Religious Support

A-33. The three broad functions of religious support include nurturing the living, caring for the wounded, and honoring the dead. Several other aspects of religious support provided by the unit ministry team include—

- Facilitating individual freedom of worship and observation of holy days according to Army regulations and mission requirements.
- Advising the commander on morals and morale as affected by religion and the impact of indigenous religions.
- Advising the commander on the ethical impact of command decisions, policies, and procedures.
- Resolving medical treatment, religious and ethical issues, religious apparel issues, and religious dietary restrictions in accordance with Army Regulation (AR) 600-20.
- Respecting the constitutional, statutory, and regulatory requirements ensuring freedom of religion for every Soldier, family member, and authorized civilian.

A-34. The unit ministry team is a task-organized team designed to support the religious, spiritual, and ethical needs of Soldiers and their families, members of other Services, and authorized civilians. The corps chaplain section advises the corps commanding general and supports the full corps by—

- Giving guidance from the commanding general in coordination with other staff.
- Establishing links with representatives of joint, multinational, interagency, faith-based organizations and religious leaders of the host nation.
- Planning and executing religious support for corps operations.
- Monitoring religious support in major subordinate commands.
- Executing support operations to sustain subordinate Army forces.

A-35. Additional chaplain resources may provide direct support and general support to the corps and other unit ministry teams depending on the mission and where assigned. See FM 1-05.

Band Support

A-36. The corps commanding general determines what musical assets are necessary in the corps AO. Bands are designed with the flexibility to employ musical performance teams in support of military operations. The corps assistant chief of staff for sustainment is the coordinating staff element responsible for band operations.

LOGISTICS

A-37. Logistics is the military art and science of carrying out the movement and maintenance of forces. Logistics includes maintenance, transportation, supply, field services, distribution, operational contract support, and general engineering support.

A-38. The corps main CP sustainment cell provides oversight for corps logistic operations. Major responsibilities include—

- Developing the corps operation plan service support annex.
- Coordinating external logistic support.
- Formulating policy, procedures, and directives related to materiel readiness.
- Formulating and implementing policy and procedures for classes of supply and related services.
- Monitoring and reporting the status of corps logistic automated information systems.
- Coordinating with internal and external activities and agencies regarding mobility operations.
- Monitoring corps logistic operations.

Maintenance

A-39. The Army uses a two-level maintenance system: field maintenance and sustainment maintenance. Field maintenance is repair and return to user. Field maintenance relies upon line replaceable unit and component replacement, battle damage assessment and repair, recovery, and services to return end items to a serviceable condition. Sustainment maintenance is repair and return to supply system. See FM 4-30.3.

A-40. TSC field maintenance activities involve collecting and analyzing maintenance data and reports. Such activities enable the TSC to enforce Army Service component command priorities relating to the repair of specific types of equipment or support of specific units. These same activities provide the means to identify significant trends and deviations from established standards. Hence, TSC maintenance managers can take action to ensure the maximum number of combat systems remain fully mission capable. TSC actions may include disseminating technical information and allocating or reallocating resources and capabilities to support maintenance requirements.

Transportation

A-41. Corps transportation requirements beyond organic lift capabilities are supported by the TSC and ESC. Collaborative planning enables units to use transportation assets efficiently and to move supplies, personnel, equipment, and units in support of corps operations. Movement throughout the theater is controlled by the TSC movement control battalion and its subordinate movement control teams. See FM 55-1.

A-42. Movement control teams process movement requests and arrange transport for moving personnel, equipment, and supplies. They process convoy clearance requests and special hauling permits. Movement control teams coordinate with the movement control battalion for the optimal mode (air, rail, inland waterway, or highway) for unprogrammed moves. These teams commit mode operators from the sustainment brigade, the logistics civil augmentation program, multinational elements, and the host nation.

A-43. The corps main CP deals with three elements of the transportation and distribution system: mode operations (how it gets there), terminal operations (how it is received and processed), and movement control (how it moves about the corps AO). While monitoring all three, the corps is most concerned with the latter. Movement control is the planning, routing, scheduling, controlling, and coordinating personnel, units, equipment, and supplies moving over multiple lines of communications. It involves synchronizing and integrating logistics efforts with other elements that span the spectrum of conflict. The corps can facilitate mission accomplishment by ensuring controlled movement of all elements. Several elements of the corps main CP focus on movement. In the main CP, the G-4's transportation element plans and monitors movement in the corps AO. The movement and maneuver cell executes terrain management for the commanding general.

Supply

A-44. Supply operations within the corps are conducted in accordance with the corps operation plan service support annex and related polices and directives. TSC directed supply and resupply actions are executed in accordance with priorities of support established by the Army Service component command. Collaboration and coordination between corps and TSC planners provides for seamless integration and synchronization with corps operations. See FM 10-27.

A-45. Typically, during the early stages of a major operation, the TSC pushes certain classes of supplies (I, IIIB, and V) to subordinate sustainment brigades and supported units. The supplies pushed stem from an analysis of the applicable supported operation plan, supported commander's priorities, and planning factors. The TSC may rely on Army pre-positioned stocks to meet initial surge requirements for sustainment. As distribution capabilities expand, a pull system is implemented to achieve greater effectiveness and efficiencies.

A-46. The TSC provides all classes of supply (less class VIII) and related services necessary to sustain Army forces throughout a major operation in the quantities and at the time and place needed. This capability includes requesting, receiving, producing, procuring, storing, protecting, relocating, and issuing the necessary supplies and services. It also includes building the necessary stockage levels in staging areas for conducting the operation and collecting, providing, and processing in-transit visibility data.

A-47. Based on parameter settings established by the TSC, the corps and theater automatic data processing service center determines if the requested item is available from within the theater and directs a materiel release order to the sustainment brigade capable of satisfying the requirement. In most instances, the processing service center automatically performs these actions in accordance with TSC-controlled parameter settings that include referral tables. Such centralized control and decentralized execution enables responsive and agile support throughout the theater, effectively minimizing customer wait time.

Field Services

A-48. The TSC plans, resources, monitors, and analyzes field services support to deployed Army forces. TSC field services operations include field laundry, showers, light textile repair, force provider, mortuary affairs, aerial delivery support, and coordination with Defense Logistics Agency for hazardous waste removal. FMI 4-93.2 discusses the field services support in detail.

Distribution

A-49. The Army distribution system is designed to optimize available infrastructure, reduce response time, maximize throughput, and support time-definite delivery. Effective distribution management synchronizes and optimizes the various subelements of the distribution system. Methods may include, but are not limited to—

- Maximizing containerization.
- Increasing standardized transportation and materials handling equipment.
- Integrating aerial resupply as a routine method of delivery.
- Synchronizing and integrating retrograde operations across all available transportation modes.
- Reducing storage.
- Reducing transportation mode transfer handling requirements.
- Increasing in-transit visibility in an AO or joint operations area (JOA).

A-50. The TSC is the distribution manager of the intra-theater segment of the global distribution system. If an ESC is deployed, it performs the role of distribution manager for its specified theater of operations or JOA. The ESC and sustainment brigades monitor, track, and execute distribution operations in accordance with TSC guidance. TSC distribution managers conduct parallel and collaborative planning with the corps headquarters to help effectively execute distribution operations throughout the corps AO.

A-51. TSC distribution managers—

- Synchronize materiel and movement management operations by maintaining logistics situational understanding through a common operational picture.
- Ensure visibility of theater distribution assets, including international organization for standardization shipping containers, aerial delivery platforms, and palletized loading system flatracks.
- Enforce established theater priorities established by the TSC or the Army Service component command.
- Maintain continuous liaison with the corps to ensure the uninterrupted flow of materiel, units, personnel, mail, and other goods.
- Synchronize retrograde support operations with an established return priority of international organization for standardization shipping containers, aerial delivery platforms, and flatracks to the distribution system.
- Coordinate directly with the theater aviation command or designated theater aviation brigades G-3 or operations staff officer (S-3) to move commodities via rotary- or fixed-wing aircraft.
- Advise the commander on the use of air movement to support distribution operations.

Operational Contract Support

A-52. Operational contract support provides additional sources of support for required supplies and services. Because of the cost of repair, complexity, system uniqueness, and maintenance capabilities, many systems are and will continue to be supported using operational contract support. The unique challenges of operational contract support require that the corps commander and staff fully understand their roles in planning for and managing contract support. Currently, three broad categories of contracted support exist:

- Theater support.
- External support.
- System support.

A-53. Theater support contracts are prearranged contracts, or contracts awarded from the mission area, by contracting officers under the command and control of the contracting support brigade or joint theater support contracting command. Contracting officers use these contracts to acquire goods, services, and minor construction support, usually from local commercial sources, to meet the immediate needs of commanders. Typically, commanders associate theater support contracts with contingency contracting. The corps headquarters often is the requiring activity (the unit requesting the support) for theater support contract support actions related to corps missions.

A-54. External support contracts provide various support to deployed forces. These contracts may be prearranged contracts or contracts awarded during the contingency itself to support the mission. Often these contracts include a mix of U.S. citizens, third country nationals, and local national subcontractor employees. The largest and most commonly used external support contract is the logistics civilian augmentation program (LOGCAP). This Army program commonly provides life support, transportation support, and other support functions to deployed Army forces and other elements of the joint force. Depending on the situation, the corps headquarters may serve as the requiring activity for major LOGCAP support requirements such as base lift support.

A-55. System support contracts are prearranged contracts by the United States Army Materiel Command life cycle management commands and separate assistant secretary of the Army (acquisition, life cycle logistics, and technology) program executive and product/project management offices. System contractors, made up mostly of U.S. citizens, provide support in garrison and may deploy with the force to both training and real-world operations. They may provide either temporary support during the initial fielding of a system (interim contracted support) or long-term support for selected materiel systems (contractor logistic support). The Army field support brigade, normally in direct support to the TSC and general support to the corps, has the lead for planning and coordinating system support contract actions. To gain an understanding of contractors on the battlefield, see FM 3-100.21.

A-56. The expeditionary contracting command field contracting support brigades, contingency contracting battalions, and senior contingency contracting teams plan and provide operational contract support for Army echelons of command from theater army through brigade. Contracting support brigades plan and provide operational contract support for Army forces operating throughout their area of operations and normally provide direct support to corps headquarters.

A-57. The contracting support brigade, normally in direct support to either ARFOR headquarters or senior sustainment command in the AO, provides the corps headquarters with general contracting planning assistance and control theater support contracting actions. The contracting support brigade staff works closely with the corps headquarters and senior sustainment command staff to ensure that the theater support contracting and LOGCAP effort is closely integrated into the overall corps sustainment effort. See FM 4-94.

A-58. For the corps headquarters, commanders ensure theater support and external contract support (primarily LOGCAP-related support) actions are properly integrated and synchronized with the overall corps sustainment effort. It is imperative the corps intelligence cell and the assistant chief of staff for sustainment work closely with the supporting sustainment command support operations, the contracting support brigade, and the supporting team LOGCAP forward. Routine corps headquarters operational contract support staff tasks include—

- **Planning.** The corps staff, with the supporting sustainment command and supporting contracting support brigade, develops applicable contract support integration plans and associated contractor management plans.
- **Developing in-theater requirements.** The corps headquarters and the command, serving as the requiring activity, prepare to develop acquisition-ready requirement packets for submission to

the supporting contracting activity. The packets include a detailed performance work statement for service requirements or detailed item descriptions for a commodity requirement. In addition to the performance work statement, these packets include an independent cost estimate and DA Form 3953 (*Purchase Request and Commitment*). Finally, the corps must be prepared to support, and possibly lead, an acquisition review board to approve and set priorities on high demand, special command interest contract support actions.

- **Assisting the contract management and contract quality control process.** In support of corps operations, the corps staff assists the contracting support brigade and team LOGCAP forward by tracking and nominating contract officer representatives. Normally these representatives are required for every service contract and LOGCAP task order. The corps headquarters and subordinate commands also need to provide receiving officials for supply contracts. Contract officer representatives and receiving officials ensure that contractors provide the contracted service or item and that this support is executed safely and effectively.

- **Assisting in contract close out.** The corps headquarters completes receiving reports. These reports certify that the Army received the contracted goods or services. The contracting officer receives a copy of the receiving report from the corps headquarters, closes the contract, and pays the contractor.

- **Participating in award fee and performance evaluation boards.** The corps headquarters or its subordinate commands often provide formal input to LOGCAP award fee and performance evaluation boards.

- **Providing contractor management oversight.** The corps commander and staff—with the theater army, contracting support brigade, team LOGCAP forward, and Army field support brigade—ensures proper contractor management execution in accordance with the contract management plan.

A-59. The corps headquarters ensures direct coordination and transfer of information related to operational contract support before transferring contracts. Additionally during unit rotations, incoming designated unit personnel actively seek out current information on contract support capabilities, policies, and procedures for their specified AO. These individuals prepare to coordinate the formal transition of existing contract management responsibilities with the redeploying unit.

A-60. Use of construction contracting and contingency funding can play an important role in support of corps operations. Civilian construction contractors and host-nation engineering support provide a significant engineering capability that becomes a force multiplier when combined with military engineering units. Construction agents often enable harnessing and directing this means of support. United States Army Corps of Engineers (USACE) support provides for technical and contract engineering support, integrating its capabilities with those of other Services and other sources of engineering-related reachback support. USACE integrates assets into the corps or theater headquarters or makes them available through a senior engineer headquarters. Whether providing construction contract and design support in the AO or outside the contingency area, USACE can obtain necessary data, research, and specialized expertise absent in theater through tele-engineering and other reachback capabilities.

General Engineering Support

A-61. General engineering requirements are coordinated with the corps movement and maneuver cell. The movement and maneuver cell recommends the allocation and employment of corps engineer assets. However, the movement and maneuver cell coordinates with the protection cell concerning general engineering support requirements related to base camp planning, development, and maintenance. General engineering support requirements beyond corps capabilities may be supported by a theater-level engineer brigade providing general support, host-nation support, or LOGCAP.

ARMY HEALTH SYSTEM SUPPORT

A-62. Army Health System support is a complex system of interrelated and interdependent systems that are designed to improve the health of Soldiers, prepare them for deployment, prevent casualties, and promptly treat injuries or illnesses that do occur. Army Health System support encompasses health service support, which supports the sustainment warfighting function and the force health protection mission, which falls under the protection warfighting function. While health service support (as a function of sustainment) is the primary focus of this publication, it is important to show the force health protection support that is also provided as part of the duties of the surgeon and members of his/her staff operating in the sustainment cell.

Health Services Support

A-63. Health service support is to the care provided to Soldiers and others with prompt treatment of wounds, injuries and illness, including behavioral illness. At the corps main CP, the corps surgeon section in the sustainment cell coordinates health service support with the modular medical units supporting the corps headquarters and its attached, OPCON, and TACON organizations. Medical activities include medical treatment, medical logistics, medical evacuation, hospitalization, dental support, preventive medicine, behavioral health, and clinical medical laboratory support. Actions of the corps surgeon and others in the sustainment cell in the main CP oversee casualty care, medical evacuation (see FM 4-02.2), and medical logistics (see FM 4-02.1).

Force Health Protection

A-64. Force health protection involves the actions taken to promote, improve, or conserve the mental and physical well-being of Soldiers. It involves identifying health threats to the force and mitigating those threats to the extent possible. The corps surgeon must stay abridged of the command and execute responsibilities in force health protection in coordination with elements of the protection functional cell. These measures protect the force from health hazards and include the prevention aspects of a number of Army Medical Department functions (preventive medicine—including food inspection, animal care missions, and prevention of zoonotic diseases transmissible to man), combat and operational stress control, dental services (preventive dentistry), and laboratory services (area medical laboratory support).

This page intentionally left blank.

Appendix B

Fires

Army doctrine identifies the fires warfighting function as the related tasks and systems that provides collective and coordinated use of Army indirect fires, joint fires, and command and control warfare, including nonlethal fires, through the targeting process. This appendix discusses fire support and fires brigade from a corps perspective.

FIRE SUPPORT

B-1. The corps headquarters has no organic fires units, but it has access to the fires battalions of its attached and operational control (OPCON) brigade combat teams (BCTs). Army fires brigades, combat aviation brigades, and other Service air and maritime fires contribute fires assets to enable the corps to accomplish its mission.

JOINT FIRE SUPPORT

B-2. *Joint fire support* is defined as joint fires that assist air, land, maritime, and special operations forces to move, maneuver, and control territory, populations, airspace, and key waters (Joint Publication (JP) 3-0). Synchronization of joint lethal fires and nonlethal fire support actions with the supported maneuver force is essential. The joint force commander (JFC) provides guidance on objectives, priorities, and desired effects.

B-3. These fires assets can be augmented with fires from land-based Marine cannon and rocket artillery and rotary- and fixed-wing assets, Air Force and Navy fixed-wing aircraft, and land- and sea-based and airborne command and control warfare systems from all Services.

ARMY FIRES

B-4. The Army is an integral part of joint fires. When deployed in support of full spectrum operations, the corps is always part of a joint force. Thoroughly understanding all aspects of joint planning and joint operations facilitates mission accomplishment. The corps commander and staff understand how to plan, develop, employ, and assess the effectiveness of joint fires.

B-5. Assets for lethal and nonlethal fires are available to the corps headquarters from the theater army. The corps headquarters synchronizes the use of Army and joint fires in support of the commander's intent by physically destroying selected enemy combat capabilities and selectively degrading or paralyzing an enemy's command and control systems through command and control warfare and other nonlethal actions. The corps commander task-organizes lethal and nonlethal assets and makes them available to the divisions and BCT assigned, attached, OPCON or under tactical control (TACON) to the corps headquarters.

B-6. The process of delivering lethal fires and nonlethal fires require two activities: integration and synchronization. Integration is the combining of fires and their effects with the other warfighting functions; synchronization is causing something to happen at the same time or in a specific time sequence. Normally commanders use fires to enable movement and maneuver; however, they can use fires separately to be decisive in an operation or to shape the fight for a follow-on decisive maneuver.

B-7. Several categories of Army fires exist: mortars, cannon, rockets and missiles, attack helicopters, and ground and airborne command and control warfare systems. Army protection fires include air defense artillery systems. Fires assets organic to the ground maneuver BCTs are supported by the joint fires and by other Service augmentation. Augmentation includes assets from the Marine Corps, Navy, and Air Force. Detachments from Marine and Navy liaison teams provide terminal guidance and observation for Marine and Navy seaborne and airborne weapons systems. The Air Force tactical air control party provides terminal attack control for close air support and air-based electronic warfare missions.

TARGETING

B-8. *Targeting* is the process of selecting and prioritizing targets and matching the appropriate response to them, considering operational requirements and capabilities (JP 3-0). The selection of targets and determination of the appropriate response to them—lethal or nonlethal fires or lethal or nonlethal actions—depends on the situation. The targeting process is an integral part of how Army headquarters uses the military decisionmaking process to solve problems (see Field Manual (FM) 5-0). As the corps commander and the main command post (CP) plans cell conduct the early steps of the military decisionmaking process, they combine the intelligence derived from the intelligence preparation of the battlefield and the running estimates to identify tasks in the targeting process as portrayed in table B-1.

Table B-1. Decide targeting process tasks

Targeting	Task
Decide	**Mission analysis.** • Develop high-value targets. • Develop targeting guidance and objectives. • Designate potential high-payoff targets. • Deconflict and coordinate high-payoff targets.
	Course of action analysis. • Develop high-payoff target list. • Establish target selection standards. • Develop attack guidance matrix. • Determine target damage assessment requirements.
	Orders production. • Finalize high-payoff target list. • Finalize target selection standards. • Finalize attack guidance matrix. • Submit info requirements and request for information to G-2 (assistant chief of staff, intelligence).

B-9. Future doctrine will describe how the fires warfighting function has several basic and sequential tasks. These tasks make up a methodology that incorporates intelligence to find the enemy and fires to kill or degrade the enemy.

CHIEF OF FIRES

B-10. The chief of fires is a coordinating staff officer serving at division to theater army level. This officer advises the commander on the best use of available fire support resources, provides input to necessary orders, and develops and implements the fire support plan. The chief of fires and the deputy chief of fires head the fires cell in the main CP and tactical CP, respectively.

B-11. The chief of fires has the fire support staff responsibility and the senior field artillery commander, normally the fires brigade commander, has a commander-to-commander relationship with the corps commander. Consequently, the relationship between the chief of fires and the organic or supporting field artillery unit commanders is analogous to the traditional relationship that has long existed between the assistant chief of staff for operations and maneuver unit commanders. The corps commander clearly delineates the personnel duties and responsibilities of the chief of fires and fires cell. The chief of fires also provides input to necessary orders and develops and implements the fire support plan. The chief of fires' primary responsibilities include—

- Developing, with the commander, chief of staff, executive officer, and assistant chief of staff, operations (G-3), a scheme of fires to support the commander's concept of operations.
- Planning and coordinating fire support tasks.
- Coordinating and synchronizing joint fires.
- Coordinating fire support assets.
- Conducting all aspects of Army indirect fires, joint fires, and command and control warfare (including nonlethal fires) through the targeting process.
- Advising the commander and staff of available fire support capabilities and limitations, as well as those of the assets available for conducting physical attack, electronic warfare, and computer network operations against enemy and adversary command and control.
- Synchronizing fire support physical attack, electronic warfare, and computer network operations.
- Leading the targeting working group and participating in the targeting meeting.
- Recommending fire support coordination measures to support current and future operations, and managing changes to those measures.
- Recommending and implementing, with the senior field artillery unit commander, the commander's counterfire plan (including radar zones), and other target engagement priorities.
- Assisting in developing the ammunition required supply rate (with the G-2, G-3, and G-4).
- Coordinating and planning for scatterable-mine use (with the engineer coordinator and supporting fires unit commander).

B-12. The corps chief of fires plans and coordinates the fires warfighting function. This officer works closely with the chief of staff and the assistant chief of staff for operations. Such cooperation ensures a clear understanding of all aspects of planning, preparation, execution, and assessment of fire support and selected aspects of command and control warfare for operations. The chief of fires assists the G-3 as needed to ensure plans flow smoothly during the transition to execution. Chapter 2 describes the corps chief of fires' primary responsibilities.

FIRE SUPPORT LIAISON AND COORDINATION

B-13. The corps and other Army commanders are supported in the use of lethal and nonlethal fires by many organizations and activities providing specialized services to the fires warfighting function. Joint force elements directly facilitate liaison and coordination with regards to fires. Liaison elements—such as the Naval and amphibious liaison element and the special operations liaison element—provide component planning and systems expertise enabling corps commanders to integrate their respective fires into the concept of operations.

B-14. Effective liaison is as important in stability and civil support operations as it is in offensive and defensive operations. This particularly applies in fire support when commanders have to communicate the effects of lethal and nonlethal fires on the target area to entities outside of the fire support system. Host nations, intergovernmental organizations, and nongovernmental organizations must understand what is happening and how it can impact their operations.

B-15. The corps command and control of fires is assisted by several organizations that provide specific joint fires expertise. These include the corps fires cell, joint targeting coordination board, and battlefield coordination detachment.

THEATER ARMY FIRES CELL

B-16. The theater army fires cell plans, coordinates, synchronizes, and executes operational joint fire support for the theater army. This cell works with the Army battlefield coordination detachment to identify the requirements for air, maritime, and special operations fire support from the appropriate joint headquarters. Coordination aims to provide the optimum lethal and nonlethal fires to support the theater commander.

JOINT TARGETING COORDINATION BOARD

B-17. The corps can serve as a joint force land component command or a joint task force headquarters. When serving in this role, the commander typically organizes a joint targeting coordination board. The joint targeting coordination board's focus is to develop broad targeting priorities and other guidance in accordance with the commander's operational objectives. The joint targeting coordination board may be an integrating center for targeting oversight efforts or a mechanism for a JFC-level review. It must be a joint activity comprised of representatives from the joint force staff, all components, and, if deemed necessary, other agencies, multinational partners, and subordinate units. The joint targeting coordination board normally facilitates and coordinates joint force targeting activities with the components' schemes of maneuver to ensure activities meet the JFC's priorities. Generally, direct coordination between elements of the joint force resolves targeting issues below the level of the joint targeting coordination board, but the joint targeting coordination board or JFC may address specific target issues not previously resolved. It addresses broad targeting oversight, coordinates targeting information, provides targeting guidance, and refines the joint integrated prioritized target list for the commander's review. See JP 3-30.

BATTLEFIELD COORDINATION DETACHMENT

B-18. Per the global force management document, each battlefield coordination detachment is assigned to an Army Service component command, which is assigned to a specific unified command. The battlefield coordination detachment—an Army liaison to a joint or combined air operations center—performs at the operational level. Its assigned tasks include ensuring the successful integration of airpower into Army ground maneuver operations. It represents Army forces or the theater army in the joint air operations center. The battlefield coordination detachment synchronizes air operations with Army ground operations by coordinating air support and exchanging operational and intelligence information. The battlefield coordination detachment helps the ground commander weight the efforts for close air support, air interdiction, tactical air reconnaissance, command and control warfare, battle status, and special weapons employment information to the land force. For additional information on the battlefield coordination detachment, see appendix E and ATTP 3-09.13.

FIRES BRIGADE

B-19. The primary Army fires organization available to the theater, corps, and division commanders is the fires brigade. It is tailored to the division operating in the joint operations area. The Army's fires brigades are tailored for a specific theater and may be task-organized based on the situation. The fires brigades give corps and other senior commanders a headquarters with which to plan, synchronize, and execute fires. The fires brigade mission provides organic and joint fires of close support and precision strike capabilities for Army and joint forces. As depicted in figure B-1 (page B-5), the brigade has organic elements to provide command and control, target acquisition, signal support, sustainment, and multiple launch rocket systems. Because fires are never left in reserve, additional fires assets are task-organized to the brigade based on the situation.

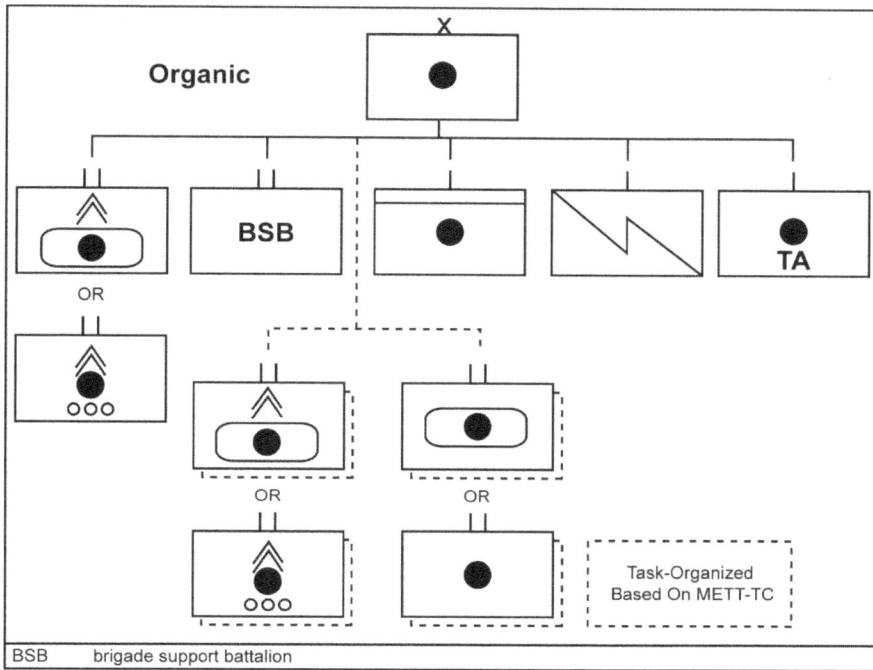

Figure B-1. Fires brigade

B-20. The fires brigade is capable of being a supported or supporting unit and providing and coordinating lethal and nonlethal fires. Fires brigades also have the necessary fire support and targeting structure to effectively execute the entire detect-decide-deliver-assess targeting process for its mission. The fires brigade can be a force field artillery headquarters capable of coordinating, synchronizing, and employing multiple Army and joint lethal and nonlethal fires units and assets in support of the force commander's concept of operation. Its capabilities include strike, counterfire, reinforcing fires for BCT organic fires, electronic attack, suppression of enemy air defense, and support of other brigades. Additionally, the fires brigade provides technical oversight of all field artillery specific training within the corps and its assigned, attached, operational control, and tactical control and supporting units.

B-21. Future doctrine discusses the capabilities of the fires brigade and fires battalion. Additional information on joint fire support coordination is found in JP 3-09.

This page intentionally left blank.

Appendix C

Corps Signal Operations

As discussed in chapter 3, effective and reliable communications allow the corps' command and control system to receive and disseminate information throughout the corps area of operations and beyond. This appendix discusses corps command and control in terms of its communications capabilities and those responsible for carrying out communications.

COMMUNICATIONS CAPABILITIES

C-1. The Army corps is a networked entity. The network links decisions to actions. The corps headquarters and its assigned organizations network with higher, lower, and adjacent military and nonmilitary formations via organic and external network capabilities.

ORGANIC NETWORK CAPABILITIES

C-2. Organic network capabilities consist of LandWarNet, tactical network support, company headquarters, and command post (CP) platoons.

LandWarNet

C-3. LandWarNet is the Army's portion of the Global Information Grid. It consists of all globally interconnected, end-to-end Army information capabilities that support warfighters, policy makers, and support personnel. LandWarNet includes all Army networks that move information that facilitates joint warfighting and supporting operations from the operational base to the edge of tactical formations, down to the individual Soldier. Paragraphs C-4 through C-17 describe how the corps' organic signal company and external signal organizations support LandWarNet capabilities for the corps headquarters.

Tactical Network Support

C-4. The corps headquarters connects, interfaces with, and draws information from the LandWarNet in garrison and when deployed. When deployed, the corps headquarters will interface with the theater army or joint task force (JTF) to gain access to required information for mission accomplishment. The Global Command and Control System (GCCS), and its Army component, GCCS-Army, are the command and control components of the Army Battle Command System employed for this purpose.

C-5. The corps headquarters tactical communications systems are mobile, deployable, quickly installed, secure, and durable. The corps headquarters network requirements include:

- Extending services from home station to the corps' area of operations (AO).
- Connecting corps CPs to higher headquarters, all Service component command headquarters, the Army Force headquarters (if the corps is not the ARFOR), and special operations forces.
- Establishing communications with forces assigned, attached, under operational control (OPCON), or under tactical control (TACON) of the corps headquarters.

C-6. In addition, the corps assistant chief of staff, signal (G-6) must:

- Coordinate for resources to meet the communication needs of lower echelon organizations without organic communications support that are attached, OPCON, or TACON to the corps headquarters.
- Provide electromagnetic spectrum management and deconfliction.

- Coordinate services for remote or non-standard users, such as multinational forces, host nations, nonstate actors, nongovernmental organizations, and others with unique communications needs.
- As required, coordinate network support for joint enablers without adequate communications.
- Support liaison operations at corps headquarters and coordinate support for corps liaison personnel at remote locations.

C-7. The corps headquarters is supported by the corps signal company (see figure C-1). The corps signal company, part of the headquarters and headquarters battalion, is configured to provide support to the corps main and tactical CPs, and the tables of organization and equipment also include the corps assistant chief of staff, signal (G-6) structure. The corps signal company is commanded by a major, who receives technical oversight from the corps G-6, a colonel.

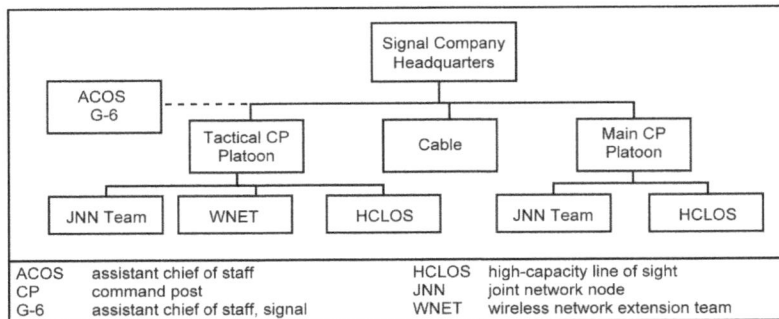

Figure C-1. The corps signal company

Company Headquarters

C-8. The corps signal company provides 24-hour support to the corps headquarters. The company consists of a headquarters, two CP support platoons, and a cable section. The platoons install, operate, maintain, and defend the communications links connecting the corps main and tactical CPs with higher, lower, and adjacent echelons.

C-9. The corps signal company headquarters provides command and control to the company. It supervises the signal elements assigned or attached to the company and provides personnel and equipment to support the company's operational mission in garrison or when deployed. It is responsible for company administrative, logistics, and maintenance support. It consists of a company headquarters and logistics support personnel (supply noncommissioned officer, armorer, and nuclear, biological, and chemical noncommissioned officer).

Command Post Platoons

C-10. The main and tactical CP platoons provide sections to install, operate, and maintain communications and connectivity to the LandWarNet for the corps' CPs. The main CP platoon has a platoon headquarters; a joint network node (JNN) section with two JNN teams and two secure, mobile, anti-jam, reliable tactical terminal teams (sometimes known as SMART–T teams); and a high-capacity line-of-sight section with two high-capacity line-of-sight teams. The tactical CP platoon has a platoon headquarters; a JNN section with one JNN team and one secure, mobile, anti-jam, reliable tactical terminal team; a high-capacity line-of-sight section with one high-capacity line-of-sight team; and two wireless network extension teams. The cable section supports the main and tactical CPs as required.

Joint Network Node Section

C-11. The JNN section includes JNN teams (Warfighter Information Network-Tactical Increment 1) and secure, mobile, anti-jam, reliable tactical terminal teams. The section installs and operates beyond line of sight links, secure voice (tactical and Defense Switched Network), Non-Secure Internet Protocol Router Network (known as NIPRNET), SECRET Internet Protocol Router Network (known as SIPRNET), Joint Worldwide Intelligence Communications System (Limited), Defense Red Switch Network, and video teleconferencing capabilities. The teams are capable of supporting forcible entry operations.

High-Capacity Line-of-Sight Section

C-12. The high-capacity line-of-sight section provides high-capacity multichannel radio links to augment or replace satellite links where feasible.

Wireless Network Extension Teams

C-13. Wireless network extension teams provide wireless network retransmission, early entry and en route communications, and support for unique communications systems such as Land Mobile Radio and Wideband Harris Radio.

Cable And Wire Section (Main CP Platoon Only)

C-14. The cable and wire section provides cable and wire teams to the main and tactical CPs as required to install, troubleshoot, repair, and evacuate cable and wire equipment. Cable and wire teams are capable of performing outside plant cable functions once the corps cable infrastructure is established.

EXTERNAL NETWORK CAPABILITIES

C-15. Because the Army corps headquarters can operate at any point across the spectrum of conflict, it may require access to signal assets from theater level and above. When needed, the corps' organic communications equipment may be augmented by systems from a theater signal command or by a national or strategic asset such as the Joint Communications Support Element. The theater army's signal organization—either a signal command (theater) or signal brigade—provides communications and information systems support to the theater army headquarters, to theater army subordinate units, and, as required, to joint and coalition organizations throughout the supported geographic combatant commander's area of responsibility. The signal command (theater) commander can assume the roles and responsibilities as the senior signal officer in the theater to include acting as the communications system directorate of a joint staff (J-6) or G-6 of the joint task force headquarters or senior Army command.

C-16. The theater signal command or brigade exercises command and control over signal organizations that have been allocated to the theater. These may include theater tactical signal brigades, subordinate expeditionary signal battalions, theater and subordinate strategic signal battalions, the theater network operations and security center, combat camera teams, and a tactical installation and networking company (see Field Manual Interim (FMI) 6-02.45). The theater signal organizations install, operate, and defend the Army portion of the joint interdependent theater network, including the theater network service center. The theater network service center provides connection, information services, and network operations capabilities. These capabilities enable the corps headquarters and units allocated to the corps to interface with both senior and subordinate units and the Global Information Grid. Currently, both the corps headquarters and units allocated to the corps may also connect to the network through a deployed division tactical hub node.

C-17. Units allocated to the corps that have no organic signal organization may receive dedicated support from an expeditionary signal battalion, or they may be located at a site that permits sharing of existing network support. Exact support requirements and allocations of pooled network resources are normally determined during the deployment planning process. Following deployment, requests for additional signal support are coordinated by the G-6 through the corps G-3.

COMMUNICATIONS RESPONSIBILITES

C-18. Responsibilities for communications falls to the corps assistant chief of staff, signal (G-6) and corps headquarters G-6 section. The latter consists of the main CP G-6 section and tactical CP signal systems.

CORPS ASSISTANT CHIEF OF STAFF, SIGNAL (G-6)

C-19. The G-6 provides the plans, operations, staff oversight, and coordination for information systems and communications to the corps headquarters and attached, OPCON, and TACON units. The G-6 is the senior signal officer in the corps and coordinates with peer counterparts at lower, adjacent, and higher echelons of command to ensure adequate network support. Should the corps headquarters serve as the core element for a joint task force, the corps assistant chief of staff, signal will become the joint task force J-6 unless superseded by a more senior signal officer.

C-20. The corps headquarters G-6 section manages the extension of Defense Switched Network and LandWarNet services throughout the corps AO. It also integrates LandWarNet assets, including strategic and tactical signal and network operations organizations, in support of corps operations. The corps G-6 coordinates with the G-3 to obtain theater resources when network missions exceed the capability of the organic signal units.

C-21. The corps G-6 section is responsible for network operations and information management functions within the corps AO. The section provides advice, direction, and guidance concerning network operations. The G-6 develops the corps network architecture and is responsible for LandWarNet operations within the corps AO, to include support of subordinate elements, if required. The G-6 section consists of elements to support the main and tactical CPs. These elements are not static and can be tailored to suit the situation.

MAIN COMMAND POST G-6 SECTION

C-22. The main CP is the corps' primary command and control element, and as such has the largest signal representation. The G-6 section manages connectivity to audio, video, written, and data systems supporting corps staff elements operating in the main CP. The corps G-6 is the senior signal officer in the corps and is chief of the signals section. Signal Soldiers are assigned to the corps signal company of the headquarters and headquarters battalion, with duty in the main CP. The main CP G-6 section—

- Monitors, manages, and controls organic communications systems that interface with the Global Information Grid.
- Plans signal support for current and future operations.
- Manages installation and operation of the main and tactical CPs local area networks and operates the corps help desk.
- Assists units allocated to the corps with network installation and troubleshooting.
- Serves as the corps information service support office.

C-23. The subordinate elements of the G-6 section are organized as shown in figure C-2.

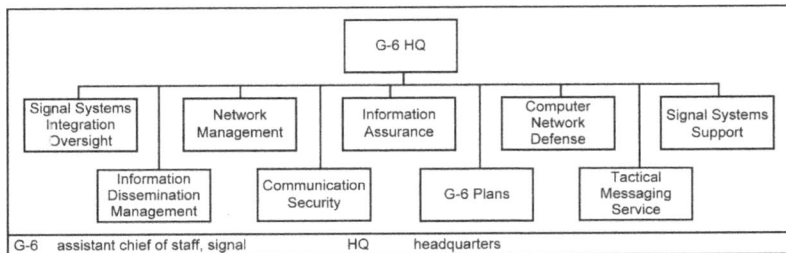

Figure C-2. Corps main command post G-6 section

TACTICAL COMMAND POST SIGNAL SYSTEMS SUPPORT ELEMENT

C-24. The tactical CP signal systems support element performs functions in the tactical CP similar to those performed by the assistant chief of staff, signal element in the main CP. The element of 26 Soldiers can be tailored (augmented or reduced) by the G-6 to meet specific mission requirements. When the tactical CP deploys, the tactical CP signal systems support element manages the local equipment and facilities that collect, protect, process, store, display, and disseminate information in the tactical CP. These Soldiers monitor, manage, and control organic communications systems that interface with the Global Information Grid, and manage a set of integrated applications, processes, and services that provide the capability for corps tactical CP staff elements to locate, retrieve, send, and receive information. Network operations functions supporting the tactical CP are normally performed by the corps network operations and security center located at the main CP.

This page intentionally left blank.

Appendix D

Airspace Command and Control

A corps commander synchronizes forces and warfighting functions in the vertical dimension in near-real time. Friendly surface, subsurface-, and air-launched weapon systems from multiple components share the airspace above the corps area of operations without hindering the application of combat power. The conduct of airspace command and control is a major responsibility. To accomplish this mission, corps commanders routinely coordinate airspace requirements with the joint, multinational, and nonmilitary airspace users. This appendix discusses the application of airspace command and control and its elements to corps headquarters operations. This appendix also discusses the responsibilities and connectivity of airspace command and control. See Field Manual 3-52 for details on airspace command and control.

AIRSPACE COMMAND AND CONTROL APPLICATION

D-1. All airspace is joint. Each joint operations area has unique airspace control requirements for its various airspace users: joint, multinational, nonmilitary, and Army (air and missile defense, Army aviation, unmanned aircraft, and field artillery). Airspace command and control focuses on integrating airspace used by this diverse set of airspace users.

D-2. Airspace control includes identifying, coordinating, integrating, and regulating airspace to increase operational effectiveness. Air Force Doctrine Document (AFDD) 2-1.7 explains that airspace control is essential to combat effectiveness in accomplishing the joint force commander's (JFC's) objectives at all levels of conflict. Marine Corps Warfighting Publication (MCWP) 3-25 states that airspace management is used to optimize the use of available airspace to allow maximum freedom, consistent with the degree of operational risk acceptable to the commander. Finally, Field Manual (FM) 3-52 reminds the reader that airspace control is provided to prevent fratricide, enhance air defense operations, and permit greater flexibility of operations. (Joint Publication (JP) 3-52 discusses joint doctrine on airspace command and control.)

D-3. Joint forces use airspace to conduct air operations, deliver fires, provide air defense, and facilitate intelligence operations. The inherent multi-Service and multinational character of airspace operations are part of an overall theater air ground system. The theater air ground system community—

- Establishes close liaison and coordination among all airspace users to facilitate unity of effort.
- Maintains common airspace control procedures implemented in an uncomplicated manner.
- Emphasizes flexibility and simplicity to retain the ability to respond to evolving enemy threat conditions and evolving friendly operations.
- Supports 24-hour and adverse weather operations.
- Strives for fratricide reduction and risk balance.
- Uses durable, reliable, redundant, and secure networks and intelligence, surveillance, and reconnaissance systems for airspace command and control.
- Ensures members of the airspace command and control team—combined arms headquarters, combat air traffic controllers, airfield operations, tactical air control parties (TACPs), air crews, airspace planners, fire support coordinators, air defenders—train as they will fight.

D-4. Airspace command and control is a command and control warfighting function task. It integrates all joint airspace users as they plan and execute the commander's intent, vision, priorities, and acceptable level of risk that maximize all airspace user capabilities and minimize adverse impacts. Airspace command and control contributes to situational understanding, enhances the common operational picture (COP), fosters coordination with organic and outside organizations, and communicates with all affected organizations. Airspace command and control employs established, agreed-upon doctrine and procedures.

D-5. Airspace command and control integrates airspace use just as other corps staff elements integrate terrain use. The corps main and tactical command posts (CPs) contain movement and maneuver cells. The main CP airspace command and control section and tactical CP airspace command and control element implement the commander's guidance as it affects airspace using the airspace command and control annex to the operation plan or operation order. They also implement the commander's priorities in the corps input to the airspace control order. Sometimes conflicts arise between requirements of different airspace users or when commander's risk guidance is exceeded. In these cases, the airspace command and control section or element attempts to integrate the requirements by modifying planned airspace use without degrading the mission effectiveness of any airspace user. If the airspace conflict cannot be resolved without degrading the mission effectiveness of an airspace user, or if the risk still exceeds risk guidance, the airspace command and control section or element deconflicts airspace use based on the commander's priorities and seeks a decision from the assistant chief of staff for operations.

D-6. Army airspace command and control doctrine does not denote that any airspace contiguous to the battlefield, or any other geographical dimension of airspace, is designated as Army airspace. Nor does it imply command of any asset that is not attached to or under operational control (OPCON) or tactical control of an Army commander. Under joint doctrine, airspace is not owned in the sense that assignment of an area of operations confers ownership of the ground. Airspace is used by multiple components, and the JFC designates an airspace control authority—usually the joint force air component commander (JFACC)—to manage airspace in the joint operations area. Even the JFC has varying degrees of control of the airspace. The commander's control depends on the characteristics of the area of operations (AO) and the political and international agreements with the host nation. Therefore, for each operation (or phase of the operation), the JFC will have more or less authority in controlling the airspace. Airspace use is negotiated as the limits of the JFC's authority changes and priorities shift.

D-7. Airspace, like ground space, is not an unlimited resource. The airspace over a corps AO is constantly in use by multiple users and can easily become saturated. Two key functions of the airspace command and control are to identify to the commander and senior staff when airspace is approaching saturation and to make recommendations for the most effective use of the airspace with the associated risk or benefit.

AIRSPACE COMMAND AND CONTROL ELEMENTS

D-8. Airspace command and control staff are organic to Army forces, brigade and higher. Corps and division both contain an airspace command and control section in their main and an airspace command and control element in their tactical CPs. The brigade combat teams (BCTs) and support brigades (except sustainment) contain a version of an air defense airspace management/brigade aviation element (ADAM/BAE). These elements integrate brigade airspace command and control, including air and missile defense and aviation functions. Each element coordinates with higher, subordinate, and adjacent elements to maximize the efficiency of airspace management and the lethality of weapon systems occupying or transiting the airspace.

THEATER-LEVEL AIRSPACE COMMAND AND CONTROL

D-9. The airspace command and control section at theater army plans and organizes the theater-level airspace command and control architecture, establishing standards and policy, publishing the airspace command and control annex, and providing the Army's input to the theater's airspace control plan, the airspace control order, and special instructions. The airspace control plan—the joint document approved by the JFC—provides specific planning guidance and procedures for the airspace control system for the joint operations area. Theater army airspace command and control planners ensure these documents adequately address subordinate units' airspace requirements. An excessively restrictive airspace control plan can hinder operations conducted by theater army subordinate units.

D-10. Most airspace command and control coordination among the joint force land component commander (JFLCC), the JFACC, the joint force maritime component commander, and other senior headquarters occurs at the joint air operations center. The JFC designates the airspace control authority and defines the relationship between it and component commanders. Normally, the JFC designates the JFACC as the airspace control authority. However, regardless of who is designated as the airspace control authority, the airspace control authority does not have the authority to approve, disapprove, or deny combat operations. That authority is only vested in operational commanders. The Army battlefield coordination detachment (BCD) serves as the ARFOR or theater army liaison to the JFACC in the joint air operations center. At this level of command, the BCD provides an airspace command and control interface between the theater airspace information systems and subordinate Army and other Service elements executing airspace command and control functions.

CORPS-LEVEL AIRSPACE COMMAND AND CONTROL

D-11. The corps headquarters oversees airspace command and control policy and standardization of tactics, techniques, and procedures throughout the corps AO. The corps airspace command and control sections in the main and tactical CPs enable this standardization by integrating all airspace requirements for the corps staff and subordinate units. The corps airspace command and control section links to the theater army airspace command and control section with the BCD to ensure that the airspace control authority planning and execution documents and policies account for corps requirements and issues.

D-12. The corps airspace command and control section is designed to execute airspace command and control even if the corps serves as an intermediate tactical headquarters, an ARFOR, a joint force land component headquarters, or a joint task force (JTF) headquarters. Airspace command and control personnel in the main and tactical CPs integrate airspace operations with the functional cells and with the integration cells. The airspace command and control element also coordinates with the tactical air control party (TACP) and the air support operations center (ASOC) colocated with the corps headquarters.

D-13. As the airspace command and control functional lead for the corps staff, the airspace command and control section develop standing operating procedures and airspace command and control annexes that help standardize airspace command and control operations among subordinate units. These procedures and annexes ensure consistency with joint airspace procedures and the theater airspace control plan, Aeronautical Information Publication, and associated plans and orders. Airspace command and control sections in the main CP perform the following functions in support of the corps mission:

- Provide airspace management expertise for the corps AO.
- Monitor joint airspace operations.
- Plan and update input to the joint airspace control plan.
- Integrate the corps airspace command and control architecture into the joint airspace command and control architecture.
- Develop the airspace control architecture to support corps plans.
- Draft all airspace command and control input for operation orders, operation plans, annexes, and estimates.
- Plan and request immediate airspace coordinating measures (ACMs).
- Deconflict airspace through appropriate authority.
- Coordinate with the corps movement and maneuver (for aviation), intelligence (for intelligence, surveillance, and reconnaissance), and fires and protection (for air and missile defense) cells.
- Provide air traffic service expertise to the corps headquarters.

D-14. The corps can be a tactical headquarters subordinate to a theater army functioning as a joint force land component or JTF. In this case, the airspace command and control section provides airspace requirements to the higher headquarters' airspace command and control section for integration into their daily airspace requests. This integration applies to the next airspace control order and for inclusion into the higher headquarters' airspace command and control annex.

D-15. During the execution phase of tactical operations, the corps headquarters normally decentralizes airspace integration to subordinate divisions and BCTs within their respective AOs. It also authorizes direct liaison between them and other theater air ground system execution airspace control nodes provided by other Services. These entities include Air Force control and reporting centers and Airborne Warning and

Control System (AWACS), Marine Corps direct air support center and tactical air operations center, and other airspace command and control entities for rapid resolution of airspace issues. For corps assigned, attached, OPCON, or tactical control BCTs or other brigades assigned their own AO, the corps delegates control over Army airspace users within the respective AOs while corps retains responsibility for integrating joint, multinational, and nonmilitary airspace users. The corps integrates all airspace requirements for corps BCTs and other brigades not assigned an AO. The corps airspace command and control section retains responsibility for airspace control over portions of the AO not assigned to subordinate units. However, even when authorizing direct liaison to subordinate units, corps retains responsibility for policy. The corps may have OPCON of a Marine air-ground task force (MAGTF). How a MAGTF integrated with corps airspace command and control depends on the size and capabilities of the MAGTF. The MAGTF's aviation combat element includes Marine air command and control system capabilities tailored for the size of the aviation combat element. Smaller regimental-based MAGTFs (with unmanned aircraft systems) may integrate in a similar manner to BCTs. Larger MAGTFs bring the full joint capability to control airspace over the MAGTF AO. Large MAGTFs can include a division-based Marine expeditionary force with the full range of Marine rotary- and fixed-wing aviation as well as a robust Marine air command and control system. A joint doctrinal relationship exists between the JFACC and JFLCC. In this case, the MAGTF requires authorized direct liaison to coordinate airspace and air operations directly with the joint air operations center.

D-16. The corps headquarters can provide airspace command and control support to multinational forces OPCON to the corps. These forces may lack airspace control capabilities and will require assistance from the corps airspace command and control section. They can be supported in a manner similar to Army functional brigades working directly for the corps.

D-17. The corps can function as a joint force land component headquarters or JTF headquarters with appropriate augmentation. (See chapter 5.) As the joint force land component headquarters, the corps airspace command and control section integrates airspace requirements among a wide range of airspace users (Army, joint, and multinational). As a JTF headquarters, the airspace command and control section works with the airspace control authority and the components to build a responsive airspace control structure. In both cases, the corps airspace command and control section develops joint force land component or JTF input to the airspace control authority through the BCD for the airspace control order, the airspace control plan, and associated plans and orders.

DIVISION AIRSPACE COMMAND AND CONTROL

D-18. Division airspace command and control oversees airspace command and control for the entire division AO, regardless of whether the AO has been further assigned to the BCT. When a division allocates part of its AO to a BCT, some airspace command and control responsibilities are delegated to the BCT. Division airspace command and control still integrates joint, multinational, and nonmilitary airspace users over the entire division AO both in planning and in execution. If the division has an unusually large AO or if the division AO is noncontiguous, then the division can delegate more airspace command and control responsibilities to the BCTs, but this may require augmentation of additional airspace command and control personnel to the BCT.

CORPS AIRSPACE COMMAND AND CONTROL ELEMENT ORGANIZATION

D-19. The airspace command and control element consists of air traffic services and air defense artillery personnel. The teaming of these Soldiers is the key to the flexibility of the airspace command and control element. Their complementary skills help the airspace command and control element perform the full range of its functions (integration, identification, coordination, regulation) necessary to control Army users of airspace. The corps airspace command and control element has personnel in the main CP's movement and maneuver cell, current operations integration cell, and in the tactical CP, if deployed. While both the main and the tactical CPs have airspace command and control personnel, only the main CP can perform all airspace command and control tasks without augmentation.

MAIN COMMAND POST AIRSPACE COMMAND AND CONTROL SECTION

D-20. The main CP's airspace command and control section has personnel in both the current operations integration cell and movement and maneuver cell. The main CP airspace command and control section can perform all airspace command and control tasks required for a tactical or operational headquarters. The corps airspace command and control section has the digital compatibility to interface with the Marine Corps and multinational automation systems (if the multinational forces receive a digital liaison officer team). The airspace command and control section lacks the capability to function as an airspace control authority. (See FM 3-52, appendix B.)

CURRENT OPERATIONS INTEGRATION CELL AIRSPACE COMMAND AND CONTROL

D-21. The current operations integration cell airspace command and control support section consists of the corps airspace command and control officer and an airspace command and control noncommissioned officer. The current operations integration cell airspace command and control support section ensures the current operations integration cell has the latest available airspace information. This support section presents airspace issues, airspace risk decisions, and airspace benefit decisions to the chief of the current operations integration cell for decisions.

MOVEMENT AND MANEUVER CELL AIRSPACE COMMAND AND CONTROL

D-22. The movement and maneuver cell airspace command and control section coordinates airspace users across warfighting functions, produces the corps' unit airspace plan, coordinates with higher airspace command and control sections, and produces the corps' input to the airspace control order. The airspace command and control section coordinates with the air and missile defense element to integrate the air and missile defense plan and air picture with the fires cell for integrated fires and with the TACP or ASOC for integrated JFACC airspace users. The main CP's airspace command and control section routinely works with the air battle manager from the ASOC if an ASOC is supporting the corps.

D-23. The airspace command and control section is manned, equipped, and trained to provide airspace control with volumes of airspace allocated to the corps by the airspace control authority. This could be control of a high-density airspace control zone or control of various types of restricted operations zones. However, the size of the corps AOs may make it impractical to control a corps high-density airspace control zone. Normally, corps decentralizes airspace command and control current operations to the divisions or BCTs AOs within the corps AO.

D-24. The corps can improve joint airspace coordination during the execution of operations by collaborating with the JFACC's control and reporting centers or AWACS. This collaboration improves the control and reporting center's situational awareness of ground operations and multi-Service coordination during operations. The corps positions some of its airspace command and control personnel at the JFACC's control and reporting center as one method to improve collaboration.

D-25. The airspace command and control section can send an airspace command and control planner with an airspace workstation to support the assistant chief of staff, plans (G-5) if required. It can shift personnel to the tactical airspace command and control section as required.

TACTICAL COMMAND POST AIRSPACE COMMAND AND CONTROL

D-26. The tactical CP can be deployed as a corps command and control facility under many circumstances. (See chapter 3.) The deputy airspace command and control officer or other designated airspace command and control leader is located in the tactical CP. The tactical CP's airspace command and control officer advises the commander and staff on airspace issues and provides increased situational awareness. This officer sends airspace issues identified by the tactical CP to the main CP for resolution, if it is available. Normally, the corps only designates the tactical CP as the lead airspace command and control element when the main CP must reposition or is otherwise unavailable. In this case, for the tactical CP to integrate airspace command and control for the corps, additional airspace command and control personnel augment the tactical CP.

SHARED AIRSPACE COMMAND AND CONTROL RESPONSIBILITIES

D-27. Within the corps headquarters, airspace command and control is a combined effort requiring the work of several cells and sections supervised by the commander and senior leaders. These cells and sections consist of the future operations and plans integrating cells, fires cell, movement and maneuver-aviation section, protection cell-air and missile defense section, and Air Force element.

FUTURE OPERATIONS AND PLANS INTEGRATING CELLS

D-28. The corps planning elements, future operations for short-term planning, and plans for long-range planning coordinate with the airspace command and control section to ensure the integration of air maneuver, fires, and protection airspace requirements. This input is forwarded to the BCD in the joint air operations center for input into the theater's airspace control order and the daily air tasking order.

D-29. During operations, the corps' three integrating cells in the main CP—current operations integrating cell, future operations cell, and plans cell—rely on the airspace command and control section for input to operation orders, operation plans, branches, and sequels. These inputs include analyzing the course of action during planning, appropriate ACMs to support current and future operations, and recommended changes to the theater airspace control plan.

FIRES CELL

D-30. Airspace command and control works with the fires cell to ensure that fire support coordination measures are integrated with the unit airspace plan. Although the airspace command and control section reviews and integrates the fire support overlay with other airspace requirements, fire support coordination measures are normally sent to higher headquarters through fires channels. The airspace command and control section and the fire support cell ensure the corps standing operating procedures and the respective annexes address the procedures for forwarding fire support coordination measures and associated airspace control measures for the appropriate chain of command. This parallel teamwork also applies to control measures such as joint fires areas, counterfire, restricted operations zones, and airspace coordination areas. The TACP element in the fires cell needs to review the airspace control order to ensure that ACMs do not unnecessarily interfere with fires and that the ACMs are in a format that the fires command and control system can process. If a conflict exists between the fires plan and the airspace control order, the airspace command and control element coordinates with the higher airspace command and control elements to correct or modify the appropriate ACM.

D-31. The airspace command and control element and the fires cell can integrate immediate fires with other airspace users. If the coverage and clarity of the air picture is good and the corps has communications with airspace users, the airspace command and control element and the fire support cell can work together to provide the corps the option of clearing fires based on the COP. This process requires near-real-time sensor data, communication with both fire units and airspace users, and integration of this process with the corps decisionmaking process for risk and benefit decisions. Airspace clearance may involve direct coordination with other theater air ground system airspace control elements, such as AWACS and the Marine Corps direct air support center. This joint coordination occurs at all levels controlling fires. Often the corps authorizes divisions and BCTs to coordinate directly with the theater air ground system airspace control elements for immediate fires. However, some situations will require the corps to retain joint fires or airspace coordination at corps level.

MOVEMENT AND MANEUVER AVIATION SECTION

D-32. Army aviation depends on the effective use of airspace to accomplish missions. Rotary- and fixed-wing assets—including unmanned aircraft systems—are generally conducted close to the ground and linked to ground maneuver at all echelons. The aviation section provides Army aviation mission data for manned and unmanned aircraft to the airspace command and control section. The airspace command and control section ensures that aviation airspace requirements are integrated into the corps airspace plan.

D-33. Most aviation mission planning (manned and unmanned) occurs at the combat aviation brigade. However, when necessary, the corps aviation section in the movement and maneuver cell can perform aviation mission planning. The airspace command and control section supports the aviation element by taking aviation mission planning data and building the appropriate ACM structure necessary to accomplish the mission. The airspace command and control section with air traffic service warrant officers and noncommissioned officers in the main and tactical CPs provides the aviation section with air traffic service expertise to assist with planning the use of air traffic service assets.

PROTECTION CELL AIR AND MISSILE DEFENSE SECTION

D-34. The air and missile defense section in the main CP protection cell is the lead staff element for the coordination of air and missile defense activities with the airspace command and control, fires, and other cells. The corps air and missile defense (AMD) section is the staff lead for integration into the joint tactical digital information link network and for the production of the COP. The airspace command and control section coordinates with the AMD section to ensure that the corps has the necessary sensor architecture to provide a complete and timely air operational picture. This includes planning the employment of air defense radars, weapons systems, and appropriate communications links. The AMD section provides the AMD plan to the airspace command and control section for integration into the corps unit airspace plan. Although the airspace command and control section reviews and integrates the corps air defense plan with other corps coordinating measures, normally the coordinating measures for the AMD plan are sent to higher headquarters through AMD channels.

AIR FORCE ELEMENTS

D-35. Some elements of the theater air control system are liaisons provided to the corps or theater army by the Air Force. These include the air mobility liaison officer, the TACP, and the ASOC. They function as a single entity in planning, coordinating, deconflicting, and integrating air support operations with ground elements. They work with the airspace command and control, AMD, and aviation section and fires cell of the main and tactical CPs. They also coordinate with liaisons, such as the battlefield coordination detachment, the theater army air and missile defense coordinator, and ground liaison officers. If an ASOC supports the corps, the ASOC and its subordinate TACPs provide direct support to the corps subordinate Army units. The ASOC plays a major role in airspace control in the corps AO providing Army procedural control of all close air support aircraft supporting corps operations.

AIRSPACE COMMAND AND CONTROL CONNECTIVITY

D-36. The corps airspace command and control section is linked to subordinate airspace command and control sections at the division and with the ADAM/BAE or air defense airspace management cells at the BCT or brigade. Together they form a network of airspace command and control capable of controlling airspace users in the low-to-medium altitudes over a corps AO.

D-37. The corps airspace command and control collaborates with theater airspace control system nodes such as the control and reporting center, AWACS, ASOC, and TACP. The corps airspace command and control section also can communicate to joint, multinational, and nongovernmental aircrafts. Airspace command and control can integrate with ASOC controllers and air battle managers to provide an integrated Army and Air Force airspace command and control control node. The corps airspace command and control section is capable of voice and digital airspace data exchange with Marine air command and control system airspace control elements and can receive digital airspace requests from a Marine Corps regiment if one is OPCON to a corps.

D-38. The corps main CP airspace command and control section contains a full suite of airspace workstations, ground-to-air radios, and communications equipment necessary to bring in the joint air operations picture independent of the other networks. The airspace command and control element in the tactical CP has airspace workstations but relies on the AMD element to bring in the air operations picture. Figure D-1 (page D-9) depicts corps airspace command and control connectivity.

D-39. Airspace command and control capabilities within the divisions, BCTs, and modular supporting brigades supplement but do not replace corps airspace command and control capabilities. Airspace command and control capabilities of the units allow the corps to focus on joint airspace integration within the corps AO. These capabilities also allow control of airspace users over portions of the corps AO unassigned to divisions and BCTs while delegating authority to the divisions and BCTs to integrate all Army airspace use within their AOs.

BRIGADE AIRSPACE COMMAND AND CONTROL

D-40. All the BCTs, multifunctional brigades, and support brigades (except sustainment) have a version of an organic ADAM/BAE. This staff element consists of air defense artillery and aviation personnel and does the airspace command and control integration function for the brigade in addition to its AMD and aviation functions. While other members of the brigade staff are key airspace command and control members (fire support cell, air liaison officer, and tactical unmanned aircraft system operators), the ADAM/BAE officer in charge is the airspace command and control integrator for the operations staff officer (S-3).

Note: An ADAM/BAE has additional aviation personnel, providing a larger aviation planning capability than the air defense airspace management element. Both have the same airspace command and control capabilities.

D-41. When a BCT controls an AO, the authority that the BCT has over Army airspace users is the same as the BCT's authority over ground units transiting its AO. All Army airspace users transiting a BCT AO coordinate with the BCT responsible for the AO they are transiting. The BCT's higher headquarters (division or corps) only integrate Army airspace use between BCTs if adjudication between BCTs is necessary. Often BCTs have the authority to coordinate directly with joint airspace control elements controlling airspace over the BCT (control and reporting center or AWACS) to coordinate fires or immediate airspace. In some situations, for example, in heavily used airspace or airspace with many joint and nonmilitary airspace users, the higher headquarters may withhold this authority. For certain situations, it may be necessary to request approval for a BCT to control a volume of airspace such as a high-density airspace control zone. However, if a BCT is to control airspace within a high-density airspace control zone for extended periods, the ADAM/BAE should be augmented. Augmentation can include additional air traffic service airspace command and control assets from the division or corps airspace command and control section or airspace information center of the combat aviation brigade and airfield operations battalions.

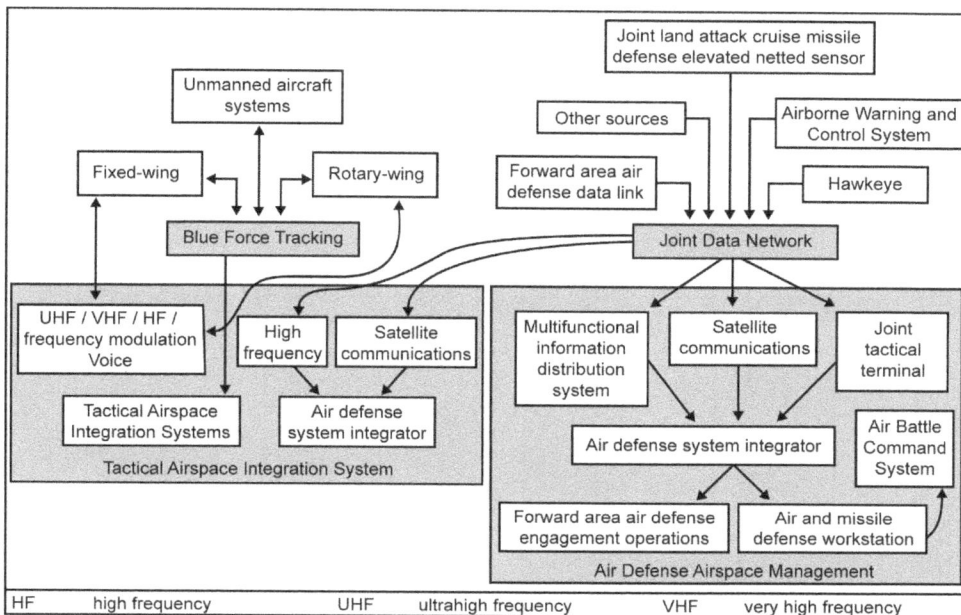

Figure D-1. Corps main command post joint airspace connectivity

D-42. Functional brigades without an organic ADAM/BAE still retain all of the brigade responsibilities for airspace command and control but rely on their higher headquarters for airspace command and control integration. If a functional brigade is under the control of a support brigade (for example, military police brigade under a maneuver enhancement brigade), the support brigade integrates the functional brigade airspace command and control requirements. If the functional brigade is directly under the control of a corps, then the corps airspace command and control section integrates the brigade airspace command and control requirements.

D-43. Several multifunctional support brigades such as the combat aviation brigade or fires brigade do not routinely control AOs but conduct operations throughout the corps AO. Normally these brigades coordinate their airspace use with the divisions and BCTs whose AOs they will transit (or with corps airspace command and control for portions of the corps AO unassigned to a division or BCT). Airspace command and control becomes more complex when corps tasks these brigades to execute a mission (such as interdiction attack or strike) that affects airspace use by other divisions or brigades. The brigade conducting the operation is the lead airspace command and control planner with the higher headquarters airspace command and control section providing planning and airspace integration support to the brigade air defense airspace management. The division or corps airspace command and control section checks that its airspace plan is adjusted to take into account the brigade commander's priorities and concept of operations.

This page intentionally left blank.

Appendix E

Air Force Interface in Corps Operations

Air support is vital to the conduct of successful corps operations. Using ground-, air-, and space-based capabilities, the Air Force supports corps operations. This appendix focuses on the Air Force airpower support to the Army. It describes the organizations involved in coordinating air support with corps headquarters. It briefly discusses each air function in support of corps headquarters and introduces common airspace coordinating measures.

INTRODUCTION

E-1. The Army corps primarily serves as an intermediate tactical headquarters under a land component command with assigned, attached, operational control (OPCON), or tactical control (TACON) of multiple divisions, including multinational or Marine Corps formations. The theater army tailors forces for the corps headquarters to meet mission requirements. As a part of the joint force, Air Force support is available to facilitate corps operations. Air Force support includes close air support (CAS); intelligence, surveillance, and reconnaissance (ISR); air interdiction; information engagement; command and control warfare; counterair; and airlift missions. The Air Force systems enhance how the Army conducts strike or attack operations using fires brigades, combat aviation brigades, or other assigned, attached, OPCON, or TACON formations.

E-2. The Air Forces' inherent ability to mass airpower—with other lethal and nonlethal fires, and maneuver at the desired place and time—makes it an important component of the operations process. Massing airpower can involve ground maneuver shaping operations to cause enemy forces to mass and become more vulnerable to air attack. Massing airpower can also support a decisive operation to attack massed enemy forces so ground forces can destroy the enemy in pieces. Air Force support is conducted at any location within the corps' area of operations (AO), including those areas assigned to subordinate divisions, brigade combat teams (BCTs), and functional brigades. Air Force ISR, command and control warfare, CAS, air interdiction, counterair, suppression of enemy air defenses, and space capabilities are important components of corps operations in unassigned areas. Those areas of the corps AO have not been further sub-allocated to subordinate organizations. (Chapter 4 discusses unassigned areas.)

E-3. Air Force support of the corps' sustaining operations routinely involves the tactical or strategic airlifting of all classes of supplies, equipment, or augmentation to corps subordinate units. Further, counterair support to preempt or counter enemy air attacks and CAS are not usually allocated to units in the corps support area. This support may be diverted from other missions to help a maneuver enhancement brigade counter a level II or III threat.

AIR SUPPORT ELEMENTS

E-4. To achieve the necessary degree of joint coordination, the Army and Air Force provide qualified personnel to work with each others' headquarters. The supporting Air Force personnel remain under the Air Force chain of command but receive logistics support from the supported Army unit. Air Force personnel come from the theater air control system and the Army air-ground system. The theater air control system consists of the air support operations center, air component coordination element, wing operations center, airborne command and control elements, and control and reporting center. The Army air-ground system consists of personnel from the battlefield coordination detachment, tactical operations center, and tactical command post (CP). See Joint Publication (JP) 3-09.3 for a thorough discussion of these organizations and their personnel.

THEATER AIR CONTROL SYSTEM

E-5. The commander, Air Force forces exercises command and control over assigned and attached forces through the theater air control system. Theater air control system provides the commander the capability to conduct joint air operations. The commander of Air Force forces' focal point for tasking and exercising OPCON over Air Force forces allocated in support of the Army is the Air Force air support operations center (ASOC), which is subordinate to the joint air operations center. Closely related to, and interconnected with the theater air control system, is the Army air-ground system. More information on the theater air control system is found in JP 3-09.3 and JP 3-30.

ARMY AIR-GROUND SYSTEM

E-6. The Army air-ground system provides the control system for synchronizing, coordinating, and integrating air operations with the land component commander's scheme of maneuver. Some elements attached to the Army air-ground system are liaisons provided by the Air Force. These elements are the air mobility liaison officer, the tactical air control party (TACP), and the ASOC. They function as a single entity in planning, coordinating, deconflicting, and integrating the air support operations with ground elements. The principal Army agencies are CPs, fire support elements, air and missile defense elements, airspace command and control elements, and coordination and liaison elements. The latter can consist of the battlefield coordination detachment (BCD), theater army air and missile defense coordinator, and ground liaison officers.

AIR SUPPORT OPERATIONS CENTER

E-7. When the corps operates as the senior tactical Army echelon (not to be confused with the term intermediate tactical headquarters), the air component provides the corps an air support operations center. The air support operations center has five primary functions. It manages CAS assets within the supported ground commander's AO; processes CAS requests and controls the flow of CAS aircraft; deconflicts airspace coordinating measures and fire support coordination measures with aircraft; assigns and directs attack aircraft, when authorized, to the joint terminal attack controllers (JTACs); and manages the Air Force air request net and its specific tactical air direction net frequencies.

AIR COMPONENT COORDINATION ELEMENT

E-8. For large operations, the joint force air component commander (JFACC) or commander, Air Force forces establishes an air component coordination element. This element better integrates air and space operations with surface operations and provides liaison with the theater army, corps headquarters, or other Army headquarters. The air component coordination element is colocated with the joint force land component commander (JFLCC) staff. The air component coordination element is the senior Air Force element assisting the senior-level Army staff in planning air component supporting and supported requirements. The air component coordination element exchanges current intelligence, operational data, and support requirements, as well as coordinates the integration of JFACC requirements for airspace coordinating measures, joint fire support coordination measures, space support, airlift support, and CAS. The air component coordination element is organized with expertise in the following areas: plans, operations, intelligence, airspace management, space, and airlift. The air component coordination element acts as the JFACC senior liaison element to senior-level Army headquarters and can also perform many air support planning functions. The air component coordination element director and corps or division air liaison officers both report directly to the air component commander. The air liaison officer does not work for the air component coordination element.

TACTICAL AIR CONTROL PARTY

E-9. The TACP is the principal Air Force liaison element aligned with the corps and division main CPs and other subordinate Army maneuver units down to battalion and company, when required. They consist of air liaison officers and JTACs. TACPs assigned to the BCT and maneuver battalion primarily advise their respective ground commanders on the capabilities and limitations of air power and assist the ground commander in planning, requesting, and coordinating CAS. The TACP provides the primary terminal attack control of CAS in support of ground forces. By coordinating directly with Army airspace and fire support agencies, the TACP deconflicts air operations in the ground sector. The TACP can use formal and informal fire support coordination measures to prevent fratricide or synchronize air operations with surface fire support. Often the corps has a TACP assigned to each division, BCT, and battalion as well as pooled terminal attack control teams. The latter provide a flexible capability to deploy down to the company level. Each BCT is supported by a TACP. However, the support brigades (battlefield surveillance, maneuver enhancement, fires, and aviation) are supported by a brigade-level TACP—depending on their assigned mission and the applicable situation—by shifting a BCT TACP to support these brigades. The supported unit is responsible for moving TACP personnel.

FIRES CELL INTERFACE WITH AIR SUPPORT OPERATIONS CENTER AND TACTICAL AIR CONTROL PARTY

E-10. The fires functional cell links the corps main CP and the ASOC or TACP for CAS missions. This cell coordinates the airspace usage with the unit's airspace command and control and air and missile defense elements. The chief of fires, ASOC, and TACP synchronize and integrate CAS for the corps headquarters. The ASOC and the corps' air liaison officer or TACP coordinates CAS with the chief of fires and the movement and maneuver cell. The TACP is also the initial point of contact for planning the integrated use of multiple effects supplied by air in addition to fire support. Additional expertise in planning integrated air operations (the use of synchronized ISR, strike, and ground forces) exists at the joint air operations center. If Navy or Marine Corps CAS is available, the air and naval gunfire liaison company may provide the corps main CP with additional liaison.

BATTLEFIELD COORDINATION DETACHMENT

E-11. The BCD supports the battlefield functions of the ARFOR commander. The BCD is an Army liaison provided by the Army Service component commander to the air operations center or component designated by the joint force commander (JFC) to plan, coordinate, and deconflict air operations. It may establish liaison with the air operations center of any Service component. See ATTP 3-09.13. The BCD—

- Processes Army requests for air support.
- Monitors and interprets the ground battle situation for the JFACC in the joint air operations center.
- Facilitates the exchange of current intelligence and operational data.

E-12. A single Army Service component command or ARFOR may consist of several corps. It is possible for the controlling ARFOR to be designated as joint force land component command or as the joint force command. In either case, the BCD singularly represents the ARFOR interests of the joint force land component command. Normally other Service or functional components provide their own liaisons to the JFACC and JFLCC as appropriate.

E-13. In a multicorps environment, each corps provides liaison to the echelons above corps headquarters. This liaison speeds the flow of information received from the BCD to the corps staff. In a single corps operation in which the corps commander is the ARFOR commander, the BCD supports the corps headquarters and colocates with the joint air operations center. If a corps or subordinate organization conducts concurrent contingency operations, the Army Service component command tailors the BCD to support the requirements of the deployed headquarters.

E-14. The BCD communicates the ARFOR commander's decisions and interests to the JFACC. As the ARFOR commander's representative in the joint air operations center, the BCD ensures the JFACC is aware of—

- The ARFOR commander's intent.
- The scheme of effects and maneuver.
- The concept for application of ground, naval, and air assets in the ARFOR AO.

E-15. The BCD monitors and interprets the land battle for the JFACC and staff. It passes ARFOR operational data and operational support requirements from the ARFOR commander to the JFACC and participating multinational forces. These support requirements include the following:

- Close air support.
- Air interdiction and mobile air interdiction.
- Manned and unmanned reconnaissance.
- Joint suppression of enemy air defenses.
- Electronic warfare.
- Airlift requirements.

E-16. The BCD does not participate directly in the ARFOR commander's estimate or decisionmaking process. The BCD supplies information regarding all the warfighting functions to ARFOR staff during the process. The ARFOR commander may delegate decisionmaking authority for specific events or situations to the BCD commander. This authority speeds action on various functions supporting the commander's plan and must be clearly defined by the ARFOR commander. The BCD eases planning, coordination, and execution of—

- Battle command.
- Intelligence.
- Firepower.
- Airspace management.
- Air and missile defense.
- Theater missile defense (when the Army air and missile defense command is not at the joint air operations center).
- Army information tasks.
- Airlift support.

JOINT TERMINAL ATTACK CONTROLLER

E-17. The JTAC is the forward Army ground commander's CAS expert. JTACs provide the ground commander recommendations on the use of CAS and its integration with ground maneuver. They are members of TACPs and perform terminal attack control of individual CAS missions. In addition to being current and qualified to control CAS, the JTAC must—

- Know the enemy situation, selected targets, and location of friendly units, and support the unit's plans, position, and needs.
- Validate targets of opportunity.
- Advise the commander on proper use of air assets.
- Submit immediate requests for CAS.
- Control CAS with supported commander's approval.
- Perform battle damage assessment.

JOINT FIRES OBSERVER

E-18. A joint fires observer can request, adjust, and control surface-to-surface fires; provide targeting information in support of Type 2 and Type 3 CAS terminal attack controls; and perform autonomous terminal guidance operations. A Type 2 observer can see either target or attacking aircraft whereas a Type 3 observer can see neither target nor attacking aircraft. JTACs cannot be in a position to see every target on the battlefield. Trained joint fires observers work with JTACs to assist maneuver commanders with the timely planning, synchronization, and responsive execution of all joint fires. Autonomous terminal guidance operations independent of CAS require the joint fires observer to communicate directly or indirectly with the individual commanding the delivery system. The observer also requires command and control connectivity with the maneuver commander or appropriate weapons release authority. Although any military member could be required to perform CAS with unqualified controller procedures, joint fires observers are better trained and prepared to execute CAS in the absence of a JTAC. A joint fires observer adds joint warfighting capability, without circumventing the need for qualified JTACs. Joint fires observers provide the capability to exploit those opportunities that exist in the corps AO. Such trained observers can efficiently support air delivered surface-to-surface fires and facilitate targeting for the JTAC in situations that are joint CAS.

AIR LIAISON OFFICER

E-19. The air liaison officer is the senior TACP member attached to a corps headquarters or subordinate ground unit who functions as the primary advisor to the ground commander on air operations. Above the battalion level, an air liaison officer is an expert in the capabilities and limitations of air power. The air liaison officer plans and executes CAS in accordance with the ground commander's intent and guidance. The air liaison officer coordinates external requests for electronic warfare support with the corps electronic warfare officer in support of corps operations. The senior air liaison officer exercises OPCON of all Air Force personnel assigned to the unit.

AIR MOBILITY LIAISON OFFICER

E-20. Air mobility liaison officers are rated Air Force officers specially trained to advise Army and Marine units on the optimum, safe use of air mobility assets. They normally support Army units at the corps, division, BCTs, and selected brigade echelons, but may support echelons above corps.

STAFF WEATHER OFFICER

E-21. The staff weather officer supports air and ground Army units with weather and weather impact information that is vital to the military decisionmaking process, including intelligence preparation of the battlefield, coordination with higher and adjacent weather teams, and support to flight mission planning. The staff weather officer performs these and other tasks within the corps assistant chief of staff, intelligence (G-2) section under the ISR operations element.

AIR FUNCTIONS IN SUPPORT OF CORPS OPERATIONS

E-22. The commander, Air Force forces supports the JFC, the JFLCC, and the corps headquarters—including its assigned, attached, OPCON, TACON, and supporting units. This support includes counterair, counterland, terminal attack control, airlift, surveillance and reconnaissance, and weather services. In addition, the commander, Air Force forces is normally dual-hatted as the JFACC and serves as the airspace control authority and the area air defense commander.

COUNTERAIR

E-23. Counterair operations aim to gain control of the air environment to achieve air supremacy. Counterair operations protect friendly forces, ensure freedom to perform other missions, and deny that freedom to the enemy. Forces conduct these operations at a distance or so to render unnecessary detailed integrating with fires and the movement of friendly ground forces. Counterair operations are consistent with the JFC's objectives and may initially involve the highest priority of all air operations. These operations involve offensive and defensive counterair operations, including the suppression of enemy air defenses. The JFACC determines the ratio of forces assigned among these counterair operations, based on—

- JFC guidance.
- Level of enemy air threat.
- Vulnerability of friendly forces to air attack.
- Enemy air defense capability.

E-24. Offensive counterair operations are essential to gaining air superiority and should be conducted at the start of hostilities to seize the offense. They are typified by attacks against—

- Command and control facilities.
- Munitions and missile storage sites.
- Aircraft on the ground or in the air.
- Any target that contributes to the enemy's airpower capability.

E-25. Suppression of enemy air defense operations are a form of offensive counterair operations designed to neutralize, destroy, or temporarily degrade enemy air defense systems and thus detract from the enemy's airpower capabilities. These operations allow friendly aviation forces to accomplish other missions effectively without interference from enemy air defense. The corps' attached, OPCON, and TACON surface-to-surface weapons complement the efforts of joint systems. The JFACC conducts suppression of enemy air defense operations against surface-to-air defensive systems. Battalion and larger ground units plan and conduct these operations in localized areas to protect fixed- and rotary-wing aircraft. The units use available field artillery cannon and rocket systems, attack helicopters, direct fire weapons, offensive information engagement, and command and control warfare.

E-26. Defensive counterair operations detect, identify, intercept, and destroy enemy airpower attempting to attack friendly forces or penetrate friendly airspace. Initially, they may be the priority mission if the enemy has seized the initiative through surprise or friendly political constraints. Defensive counter air operations involve active measures such as using combat fighter aircraft and air defense artillery. They also involve passive measures, not involving weapons systems, such as—

- Radar coverage for early warning.
- Protective construction (for example, hardened sites).
- Cover, camouflage, deception, dispersion, and frequent movement of personnel and equipment.

See JP 3-01 for additional information on countering air and missile threats.

COUNTERLAND

E-27. Counterland is air and space operations against enemy land force capabilities to create effects that achieve JFC objectives. These operations dominate the surface environment and prevent the opponent from doing the same. Although historically associated with support to friendly land forces, counterland operations may encompass missions either without the presence of friendly land forces or with only a few land forces providing target cueing. This independent or direct attack of adversary surface operations by air and space forces is the key to success when seizing the initiative during early phases of a campaign. Counterland provides two discrete air operations for engaging enemy land forces: air interdiction and CAS. Air interdiction uses air maneuver to indirectly support land maneuver or directly support an air scheme of maneuver. CAS uses air maneuver to directly support land maneuver.

E-28. Interdiction operations are joint actions to divert, disrupt, delay, or destroy the enemy's military potential before it can be used effectively against friendly forces, or otherwise meet JFC objectives. It may—

- Reduce the enemy's capability to mount an offensive.
- Restrict the enemy's freedom of action and increase vulnerability to friendly attack.
- Prevent the enemy from countering an increase in friendly strength.
- Decrease the enemy's reserves.

Air Interdiction

E-29. Normally the JFACC executes air interdiction as part of a systematic and persistent operation in support of the JFC's intent. Air interdiction includes actions against land targets positioned to have a near-term effect on the corps' operations but still not in proximity to the corps maneuver and support forces. The corps headquarters nominates these air interdiction targets. The theater army and JFLCC prioritize corps-nominated air interdiction targets. The theater army and JFLCC priorities are submitted to the JFACC along with those of other functional or Army Service components in theater and the JFC's objectives. Air interdiction requires joint coordination during planning.

E-30. Air interdiction occurs at such distance from friendly forces that detailed integration of each air mission with the fire and movement of friendly forces is not required. When air interdiction occurs inside the fire support coordination line, the ASOC is the principal command and control node for direction and controls the missions, ensuring the necessary coordination with ground operations.

E-31. Conducting accurate and effective attacks on targets far beyond the corps' maneuver forces helps to establish the conditions necessary for the conduct of the division's decisive operation. The JFLCC may provide a portion of the sorties allocated to air interdiction to the division commander. Normally, however, the division commander may only nominate targets for the air commander to attack.

E-32. The JFACC oversees executing air interdiction operations. Air interdiction in support of the Army commander disrupts the continuity of the enemy's operations. Objectives may include—

- Reducing the enemy's capability to employ follow-on forces.
- Preventing the enemy from countering friendly maneuver.
- Hindering the enemy's ability to resupply its committed forces.

E-33. In truly joint interdependent operations, the corps commander may be the supporting commander during air interdiction operations by using friendly fire and maneuver forces to cause the enemy to mass or break cover, thus increasing the enemy's vulnerability to air attack. Although forces can nominate air interdiction targets by specific unit, time, and place of attack, describing the desired results or objectives to the air commander often proves more effective. This use of mission-type targets allows the air commander greater flexibility in planning and executing the attack. However, commanders can recommend or request specific munitions against a target that is particularly vulnerable to the munitions requested.

E-34. The corps' air interdiction targeting process does not stop with nomination of the targets or mission-type requests. Target intelligence continues from when the target nomination is made to when the unit detects and tracks the target to when the unit finally attacks the target. The corps main CP allocates intelligence and surveillance assets to support the combat assessment of targets attacked by both CAS and air interdiction. The corps headquarters and Air Force share close and continuous intelligence, particularly for targets that have limited dwell time or cannot be accurately located until just prior to attack. (See JP 3-03 for additional information on joint interdiction operations.)

Close Air Support

E-35. Close air support is an attack against hostile surface forces in proximity to friendly forces and requires detailed integration into the supported commander's scheme of fires and maneuver. To be successful, CAS responds to the ground commander's needs. CAS targets are selected by the ground commander. Elements of the theater airspace control system plan, direct, and control CAS. CAS enhances ground force operations by providing the capability to deliver many weapons and massed firepower at decisive points. CAS is conducted to—

- Blunt an enemy attack on a friendly position.
- Help obtain and maintain the ground offensive.
- Provide cover for friendly movements.

E-36. Normally the JFLCC distributes CAS to subordinate Army commanders who then redistribute their CAS distribution to their subordinate commanders. By retaining control over most of the CAS sorties, the corps and its subordinate commands can shift priorities, weight its effort, and rapidly respond to emerging opportunities without shifting CAS sorties from one BCT to another. Combining CAS with attack helicopters and artillery produces a highly effective joint air attack team.

TERMINAL ATTACK CONTROL

E-37. Recent technological advances in aircraft capabilities, weapons systems, and munitions have provided joint terminal attack controllers with additional tools to maximize effects of fires while mitigating risk of fratricide when employing air power near friendly forces. During CAS, some technologies can be exploited: aircraft and munitions equipped with a global positioning system, laser range finders and designators, and digital system capabilities. Terminal attack control procedures exploit advances in technology.

E-38. Three types of terminal attack control exist. Each type follows a set of procedures with an associated risk. Commanders consider the situation and issue guidance to the joint terminal attack controller based on recommendations from their staff and associated risks identified in the tactical risk assessment. The aim is to offer the lowest-level supported commander, within the restraints established during risk assessment, the latitude to determine which type of terminal attack control best accomplishes the mission. Specific levels of risk should not be associated with a given type of terminal attack control (for example, digital targeting systems used in Type 2 control may be a better mitigation of risk than using Type 1). The three types of control are not ordnance specific.

AIRLIFT

E-39. Airlift is the transportation of personnel and materiel through the air, which can be applied across the spectrum of conflict to achieve or support objectives. Airlift can achieve tactical and strategic effects. It provides rapid and flexible mobility options that allow civilian and military forces as well as government agencies to respond to and operate in a wider variety of circumstances and timeframes. It provides U.S. forces with the global reach to apply strategic global power quickly to crisis situations by delivering necessary forces. The power projection capability for airlift supplies is vital since it provides the flexibility to get rapid-reaction forces to the point of a crisis with minimum delay. Airlift can serve as a United States presence worldwide, demonstrating the Nation's resolve and serving as a constructive force during times of humanitarian crisis or natural disaster.

E-40. Corps requests for intertheater (between different theaters) airlift are handled by the Air Force air mobility liaison officer supporting the corps through United States Transportation Command. Corps requests for intratheater (within a theater) airlift support are handled through Army logistic channels, with variations for the immediacy of the request. See JP 3-17 and Air Force Doctrine Document (AFDD) 2-6 for additional information concerning air mobility.

SURVEILLANCE AND RECONNAISSANCE

E-41. Surveillance is the function of systematically observing air, space, surface, or subsurface areas, places, persons, or things by visual, aural, electronic, photographic, or other means. Surveillance is a continuing process not oriented to a specific target. In response to the requirements of military forces, surveillance must be designed to provide warning of enemy initiatives and threats and to detect changes in enemy activities.

E-42. Air- and space-based surveillance assets exploit elevation to detect enemy initiatives at long range. For example, its extreme elevation makes space-based, missile-launch detection and tracking indispensable for defense against ballistic missile attack. Surveillance assets are now essential to national and theater defense and to the security of air, space, subsurface, and surface forces.

E-43. Reconnaissance complements surveillance by obtaining specific information about activities and resources of an enemy or potential enemy through visual observation or other detection methods. Reconnaissance also complements surveillance by securing data concerning the meteorological, hydrographic, or geographic characteristics of a particular area. This can be an important part of the corps targeting process. Locations and activities targeted for surveillance and reconnaissance also reveal important civil considerations during operations focused on the conduct of stability operations. Reconnaissance generally has a time constraint associated with the tasking. The corps main CP normally handles preplanned requests for aerial reconnaissance; the appropriate TACP handles immediate requests. See AFDD 2-9 for additional information on this air operational function.

E-44. Combining the effects of ISR, strike, and ground assets provides a synergistic effect and maximizes the use of limited assets ensuring joint force success. Integrating air and space assets to supply multiple effects toward accomplishing operations objectives requires coordination among the corps main CP functional, integrating, and coordinating cells and staffs.

WEATHER SERVICES

E-45. Weather services conducted by the Air Force provide timely and accurate environmental information, including both space environment and atmospheric weather, to Army and joint commanders. Weather services gathers, analyzes, and provides meteorological data for mission planning and execution. Environmental information is integral to the decision process and timing for employing forces and conducting air, ground, and space launch operations. Weather services also influences the selection of targets, routes, weapons systems, and delivery tactics, and act as a key element of information superiority. See JP 3-59 and AFDD 2-9.1 for additional information on weather services.

COMMON AIRSPACE COORDINATING MEASURES

E-46. Often highly concentrated friendly surface, subsurface, and air-launched weapon systems share the airspace without hindering the application of combat power with the JFC's intent. Normally the JFC designates an airspace control authority to meet the airspace requirements of subordinate commanders. The airspace control authority assumes responsibility for operating the airspace control system and achieves unity of effort primarily through centralized planning and control. Airspace coordination primarily enhances combat effectiveness of the joint force. Basic principles of airspace coordination include the following:

- The airspace control system supports JFC objectives and facilitates unity of effort.
- Close coordination between air traffic control and air defense elements reduces the risk of friendly fires and increases the effectiveness of air defense.
- Close liaison and coordination among all airspace users inside and outside the operational area promotes timely and accurate information flow to airspace managers.
- Airspace control procedures provide maximum flexibility by effectively mixing positive and procedural control measures.
- The procedural control measures are uncomplicated and readily accessible to all forces.
- The airspace control system in the combat zone has a reliable, jam-resistant, and secure communications network.
- Air control assets of the airspace control system have built in redundancy for survivability on the battlefield.
- The structure of the airspace control system responds to developing enemy threats and the unfolding operation.
- Airspace control functions rely on airspace coordinating measures resources, but these functions are separate and distinct from real-time control of aircraft and the terminal air traffic controller environment.
- Flexibility and simplicity is emphasized throughout to maximize the effectiveness of forces operating within the system.
- Airspace control needs to support 24-hour operations in all-weather and environmental conditions.

E-47. The methods of airspace control range from positive control of all air assets in an airspace control area to procedural control of all such assets, or any effective combination of the two. Airspace control systems need to accommodate these methods based on component, joint, and national capabilities and requirements. Positive control relies on radars; other sensors; identification, friend or foe selective identification features; digital data links; and other elements of the air defense system to positively identify, track, and direct air assets. Procedural control relies on airspace coordinating measures such as—

- Comprehensive air defense identification procedures and rules of engagement.
- Low-level transit routes.
- Minimum-risk routes.
- Aircraft identification maneuvers.
- Fire support coordination measures.
- Coordinating altitudes.
- Restricted operations zones and restrictive fire areas.
- Standard use Army aircraft flight routes.
- High-density airspace control zones.

See JP 3-52, AFDD 2-1.7, and FM 3-52 for additional information concerning airspace coordinating measures.

Glossary

The glossary lists acronyms and terms with Army or joint definitions, and other selected terms. Where Army and joint definitions are different, *(Army)* follows the term. The proponent manual for terms is listed in parentheses after the definition.

SECTION I – ACRONYMS AND ABBREVIATIONS

ABCS	Army Battle Command System
ACM	airspace coordinating measure
ADAM/BAE	air defense airspace management/brigade aviation element
AFDD	Air Force doctrine document
AMD	air and missile defense
AO	area of operations
AR	Army regulation
ARFOR	*See* ARFOR under terms.
ASOC	air support operations center
AWACS	Airborne Warning and Control System
BCD	battlefield coordination detachment
BCT	brigade combat team
CAS	close air support
CBRN	chemical, biological, radiological, and nuclear
CBRNE	chemical, biological, radiological, nuclear, and high-yield explosives
CG	commanding general
CJCS Guide	Chairman of the Joint Chiefs of Staff guide
CJCSI	Chairman of the Joint Chiefs of Staff instruction
CJCSM	Chairman of the Joint Chiefs of Staff manual
CJCSN	Chairman of the Joint Chiefs of Staff notice
COP	common operational picture
CP	command post
DA	Department of the Army
ESC	expeditionary sustainment command
FM	field manual
FMI	field manual-interim
G-1	assistant chief of staff, personnel
G-2	assistant chief of staff, intelligence
G-3	assistant chief of staff, operations
G-4	assistant chief of staff, logistics
G-5	assistant chief of staff, plans
G-6	assistant chief of staff, signal
G-7	assistant chief of staff, information engagement
G-8	assistant chief of staff, financial management

G-9	assistant chief of staff, civil affairs operations
ISR	intelligence, surveillance, and reconnaissance
J-1	manpower and personnel directorate of a joint staff
J-4	logistics directorate of a joint staff
J-5	plans directorate of a joint staff
J-6	communications system directorate of a joint staff
J-7	operational plans and interoperability directorate of a joint staff
JFACC	joint force air component commander
JFC	joint force commander
JFLCC	joint force land component commander
JMD	joint manning document
JNN	joint network node
JOA	joint operations area
JP	joint publication
JTAC	joint terminal attack controller
JTF	joint task force
LOGCAP	logistics civilian augmentation program
MCWP	Marine Corps warfighting publication
NIPRNET	Non-Secure Internet Protocol Router Network
OPCON	operational control
S-3	operations staff officer
SIPRNET	SECRET Internet Protocol Router Network
SJFHQ(CE)	standing joint force headquarters (core element)
TACON	tactical control
TACP	tactical air control party
TAMD	theater air and missile defense
TSC	theater sustainment command
UAS	unmanned aircraft system
U.S.	United States

SECTION II – TERMS

ARFOR

The Army Service component headquarters for a joint task force or a joint and multinational force. (FM 3-0)

References

Field manuals and selected joint publications are listed by new number followed by old number.

REQUIRED PUBLICATIONS

These documents must be available to intended users of this publication.

FM 1-02 (101-5-1). *Operational Terms and Graphics.* 21 September 2004.

JP 1-02. *Department of Defense Dictionary of Military and Associated Terms.* 12 April 2001.

RELATED PUBLICATIONS

These documents contain relevant supplemental information.

JOINT AND DEPARTMENT OF DEFENSE PUBLICATIONS

Most joint publications are available online: http://www.dtic.mil/doctrine/new_pubs/jointpub.htm.

CJCS Guide 3501. *The Joint Training System - A Primer for Senior Leaders.* 31 July 2008.

CJCSI 1301.01C. *Individual Augmentation Procedures.* 1 January 2004.

CJCSI 3100.01B. *Joint Strategic Planning System.* 12 December 2008.

CJCSM 3500.04E. *Universal Joint Task Manual.* 25 August 2008.

CJCSN 3500.01. *2009–2010 Chairman's Joint Training Guidance.* 8 September 2008.

JP 1. *Doctrine for the Armed Forces of the United States.* 2 May 2007.

JP 1-0. *Personnel Support to Joint Operations.* 16 October 2006.

JP 2-03. *Geospatial Intelligence Support to Joint Operations.* 22 March 2007.

JP 3-0. *Joint Operations.* 17 September 2006.

JP 3-01. *Countering Air and Missile Threats.* 5 February 2007.

JP 3-03. *Joint Interdiction.* 3 May 2007.

JP 3-08. *Interagency, Intergovernmental Organization, and Nongovernmental Organization Coordination During Joint Operations* (2 volumes). 17 March 2006.

JP 3-09. *Joint Fire Support.* 30 June 2010.

JP 3-09.3. *Close Air Support.* 8 July 2009.

JP 3-16. *Multinational Operations.* 7 March 2007.

JP 3-17. *Air Mobility Operations.* 2 October 2009.

JP 3-18. *Joint Forcible Entry Operations.* 16 June 2008.

JP 3-28. *Civil Support.* 14 September 2007.

JP 3-30. *Command and Control for Joint Air Operations.* 12 January 2010.

JP 3-31. *Command and Control for Joint Land Operations.* 29 June 2010.

JP 3-50. *Personnel Recovery.* 5 January 2007.

JP 3-52. *Airspace Control.* 20 May 2010.

JP 3-59. *Meteorological and Oceanographic Operations.* 24 September 2008.

JP 3-60. *Joint Targeting.* 13 April 2007.

JP 3-63. *Detainee Operations.* 30 May 2008.

JP 4-0. *Joint Logistics*. 18 July 2008.

JP 5-0. *Joint Operation Planning*. 26 December 2006.

ARMY PUBLICATIONS

Most Army doctrinal publications are available online: <http://www.army.mil/usapa/doctrine/ Active_FM.html>. Army regulations are produced only in electronic media. Most are available online: < http://www.army.mil/usapa/epubs/index.html>.

AR 20-1. *Inspector General Activities and Procedures*. 1 February 2007.

AR 27-1. *Judge Advocate Legal Services*. 30 September 1996.

AR 165-1. *Army Chaplain Corps Activities*. 3 December 2009.

AR 220-1. *Army Unit Status Reporting and Force Registration – Consolidated Policies*. 15 April 2010.

AR 360-1. *The Army Public Affairs Program*. 15 September 2000.

AR 600-20. *Army Command Policy*. 18 March 2008.

ATTP 3-09.13. *The Battlefield Coordination Detachment*. 21 July 2010.

DA Pam 600-81. *Information Handbook for Operating Continental United States (CONUS) Replacement Centers and Individual Deployment Sites*. 15 July 2001.

DA Pam 690-47. *DA Civilian Employee Deployment Guide*. 1 November 1995.

FM 1. *The Army*. 14 June 2005.

FM 1-0. *Human Resources Support*. 6 April 2010.

FM 1-01. *Generating Force Support for Operations*. 2 April 2008.

FM 1-04. *Legal Support to the Operational Army*. 15 April 2009.

FM 1-05. *Religious Support*. 18 April 2003.

FM 1-06. *Financial Management Operations*. 21 September 2006.

FM 2-0. *Intelligence*. 23 March 2010.

FM 3-0. *Operations*. 27 February 2008.

FM 3-04.15. *Multi-Service Tactics, Techniques, and Procedures for the Tactical Employment of Unmanned Aircraft Systems*. 3 August 2006.

FM 3-04.111. *Aviation Brigades*. 7 December 2007.

FM 3-05 (100-25). *Army Special Operations Forces*. 20 September 2006.

FM 3-05.40. *Civil Affairs Operations*. 29 September 2006.

FM 3-07. *Stability Operations*. 6 October 2008.

FM 3-24. *Counterinsurgency*. 15 December 2006.

FM 3-28. *Civil Support Operations*. 20 August 2010.

FM 3-28.1. *Multi-Service Tactics, Techniques, and Procedures for Civil Support (CS) Operations*. 3 December 2007.

FM 3-34. *Engineer Operations*. 2 April 2009.

FM 3-39. *Military Police Operations*. 16 February 2010.

FM 3-50.1. *Army Personnel Recovery*. 10 August 2005.

FM 3-52. *Army Airspace Command and Control in a Combat Zone*. 1 August 2002.

FM 3-90.31. *Maneuver Enhancement Brigade Operations*. 26 February 2009.

FM 3-100.21 (100-21). *Contractors on the Battlefield*. 3 January 2003.

FM 4-0. *Sustainment*. 30 April 2009.

FM 4-02.1. *Army Medical Logistics*. 8 December 2009.

FM 4-02.2. *Medical Evacuation*. 8 May 2007.

FM 4-30.3. *Maintenance Operations and Procedures*. 28 July 2004.

FM 4-94 (FM 4-93.4). *Theater Sustainment Command*. 12 February 2010.

FM 5-0. *The Operations Process*. 26 March 2010.

FM 5-19. *Composite Risk Management*. 21 August 2006.

FM 6-0. *Mission Command: Command and Control of Army Forces*. 11 August 2003.

FM 6-02.43. *Signal Soldier's Guide*. 17 March 2009.

FM 7-0. *Training for Full Spectrum Operations*. 12 December 2008.

FM 10-27. *General Supply in Theaters of Operations*. 20 April 1993.

FM 55-1. *Transportation Operations*. 3 October 1995.

FMI 4-93.2. *The Sustainment Brigade*. 4 February 2009.

FMI 6-02.45. *Signal Support to Theater Operations*. 5 July 2007.

OTHER PUBLICATIONS

ABCA Coalition Operations Handbook. 14 April 2008.

AFDD 2-1.7. *Airspace Control in the Combat Zone*. 13 July 2005. Available at http://www.cadre.maxwell.af.mil/.

AFDD 2-6. *Air Mobility Operations*. 1 March 2006. Available at http://www.cadre.maxwell.af.mil/.

AFDD 2-9. *Intelligence, Surveillance, and Reconnaissance Operations*. 17 July 2007. Available at http://www.cadre.maxwell.af.mil/.

AFDD 2-9.1. *Weather Operations*. 3 May 2006. Available at http://www.cadre.maxwell.af.mil/.

MCWP 3-25. *Control of Aircraft and Missiles*. 26 February 1998. Available at https://www.doctrine.quantico.usmc.mil/.

United States Code, Title 32. *National Guard*. Available at http://www.gpoaccess.gov/.

WEB SITES RECOMMENDED FOR ADDITIONAL INFORMATION

Sites listed were accessed October 2010.

Combined Arms Support Command. http://www.cascom.army.mil/.

Defense Logistics Agency. http://www.dla.mil/.

Defense Threat Reduction Agency. http://www.dtra.mil/.

Joint Personnel Recovery Agency. http://www.jpra.jfcom.mil/.

Joint Public Affairs Support Element. http://www.jfcom.mil/about/abt_jpase.htm.

Joint Fires Integration and Interoperability Team. http://www.eglin.af.mil/jfiit.asp.

Joint Systems Integration Center. http://www.jfcom.mil/about/com_jsic.htm.

Joint Communications Support Element. http://www.jcse.mil/live09/index_n.htm.

Navy Warfare Development Command. http://www.navy.mil/local/nwdc/.

Navy Doctrine Library System. https://ndls.NWDC.navy.mil/.

North East Regional Response Center–Special Projects Office. http://peoc3t.monmouth.army.mil/spo/spo_djc2.html.

Sustainment Center of Excellence Battle Command Knowledge System. https://lognet.bcks.army.mil/.

Universal Joint Task List (UJTL). http://www.dtic.mil/doctrine/training/ujtl_tasks.htm

PRESCRIBED FORMS

None.

REFERENCED FORMS

DA Form 2028. *Recommended Changes to Publications and Blank Forms*.

DA Form 3953. *Purchase Request and Commitment*.

This page intentionally left blank.

Index

Entries are by paragraph number unless specified otherwise.

www.ingramcontent.com/pod-product-compliance
Lightning Source LLC
LaVergne TN
LVHW081323060426

835511LV00011B/1831